Fokker Aircraft of WWI

Volume 5 | 1918 Designs

Part 1 – Prototypes & D.VI

Jack Herris & Jörn Leckscheid

Great War Aviation Centennial Series #55A

AF822551

This book is dedicated to Eriká Imrie. Thanks for tolerating Alex's hobby for so many years, and for being such a great friend!

Acknowledgements

My sincere thanks to Achim Engels and Gerry Mos for scale drawings and photographs, Dave Watts for photographs, Greg VanWyngarden for photographs and helpful suggestions, Colin Owers for photos and documents, Lance Bronnenkant, Richard Andrews, Dave Hooper, and Bill Toohey for photographs, Reinhard Kastner, Bruno Schmäling, Dr. Hannes Täger, Tobias Weber, and Reinhard Zankl for photos and information, Bob Pearson for his color profiles, Taz Philips for valuable feedback and the cover painting, Aaron Weaver for photographs and cover design, and the Museum of Flight in Seattle and the Deutsches Technikmuseum in Berlin for photographs. Any errors are my responsibility.

Color aircraft profiles © Bob Pearson. Purchase his CD of WWI aircraft profiles for $50 US/Canadian, 40 €, or £30, airmail postage included, via Paypal to Bob at: **bpearson@kaien.net**

Cover painting by Steve Anderson.

For our aviation books, please see our website at: **www.aeronautbooks.com**.

I am looking for photographs of the less well-known German aircraft of WWI to complete this series. For questions or to help with photographs you may contact me at **jherris@me.com**.

ISBN: **978-1-953201-09-6**

Design and layout: Jack Herris
Cover design: Aaron Weaver
Digital photo editing: Jack Herris

Aeronaut Books

Books for Enthusiasts by Enthusiasts

www.aeronautbooks.com

Table of Contents

Fokker at His Most Creative	3
Fokker Aircraft Type Summary	5
Fokker V9	6
Fokker V10	9
Fokker V11	10
Fokker V12	18
Fokker V13	20
Fokker V14, V15, V16	21
Fokker V17	22
Fokker V18	26
Fokker V19	29
Fokker V20	30
Fokker V21	34
Fokker V22	38
Fokker V23	48
Fokker V24	52
Fokker V25	58
Fokker V26	61
Fokker V27	62
Fokker V28	66
Fokker V29	74
Fokker V30, V31, V32	77
Fokker V33	78
Fokker V34	82
Fokker V35	86
Fokker V36	88
Fokker V37	94
Fokker V38	100
Fokker V39	104
Fokker V40	108
Fokker V41	112
Fokker V42	114
Fokker V43	116
Fokker V44 & V45	118
Fokker D.VI	120
Bibliography	150
Afterword	174

This fifth volume of the Fokker set tells the story of the later Fokker V series of experimental airplanes and Fokker's late-war production fighters, the D.VI, D.VII, and E.V/D.VIII.

The volume was written as one part, but the Amazon KDP system could not process the file with so many images. As a result we have had to break the book into two parts to publish it.

Part 1 of this volume covers Fokker's late-war prototypes and the D.VI, 120 aircraft of which were ordered but only 60 produced because the second half of the order was later changed to produce the higher-performance E.V/D.VIII.

As authors, our task is to document the facts. In one case we have also given our opinion – clearly indicated as such – about an *Idflieg* production decision.

Cover: **Test Flight over Mecklenburg!** The skies over Schwerin and the surrounding Mecklenburg countryside must have been abuzz in late 1917 through 1918 due to the many aircraft designs being developed and test flown by Fokker test pilots. Here the V25 cantilever monoplane design is being taken for a test flight by an eager pilot. Although to modern eyes this seems like a step in the right direction design wise, it was rejected due to the forward and downward view being obstructed. The designs that were chosen for production were two high-wing parasol designs, the Fokker V28 and V29. *(Cover illustration by Aaron Weaver)*

Fokker at His Most Creative

The final year of the war was Fokker's most creative and successful of his career. His Dr.I had been reasonably successful but its extra drag and low-power rotary engine prevented it from being a complete success. Most important was his development of successful thick wooden cantilever wing technology that was first used in the Dr.I. Now he needed more powerful engines and to find the right airframe configuration.

At the Red Baron's suggestion *Idflieg* had decided to hold a fighter competition in late January 1918. With his characteristic energy and creativity, Fokker designed a number of fighters for this competition that all utilized his thick wooden wings. These were all variations of different configurations that used different engines. All Fokker fighters were fitted with, or designed for, twin synchronized machine guns.

Fokker at the First Fighter Competition

Fokker retained many of the features of his Triplane, including his wooden cantilever wings, airfoil over the axle, and construction materials and methods. The problem was the triplane configuration itself, which inherently had more drag than an equivalent biplane. After the Dr.I Triplane, all following Fokker fighter prototypes were monoplanes or biplanes using Fokker's wood cantilever wing.

For Germany's First Fighter Competition held in January 1918, Fokker submitted nine prototypes, far more than any other manufacturer. The Pfalz company submitted five fighters to secure second place in the number of aircraft entered. Fokker's aircraft dominated the competition. The V9 and V13 rotary-powered biplane prototypes that were developed into the production D.VI were the natural successors to the similarly-powered Triplane. But the higher altitudes where combats now started demanded more powerful engines, especially with more power at high altitude, and the V11 and V18 biplane prototypes, which were developed into the production D.VII, offered that power due to their use of the Mercedes D.III. Both the D.VI and D.VII benefitted from improvements to Fokker's wooden cantilever wing and their biplane configuration.

Considering the Dr.I Triplane was then serving at the front, it is interesting that Fokker entered two monoplane fighter designs with cantilever wings into the First Fighter Competition; one was rotary powered and the other had the Mercedes. Despite its low-power 110/120 hp Oberursel Ur.II rotary, the same engine used in the Dr.I Triplane, the Fokker V17 monoplane was the fastest fighter in the competition and was also very maneuverable. Notwithstanding its exceptional performance, it was not selected for production. The official reason was the mid-mounted wing restricted the pilot's downward field of view, and indeed it did. However, the successful early Fokker *Eindecker* fighters all had the same issue. The actual reason may have been lack of confidence in its monoplane configuration. Regardless, not placing a production order for the Fokker V17 seems like a mistake given the near universal use of the low-wing monoplane configuration for WW2 fighters.

Monoplane or Biplane?

In the author's opinion, the V17 monoplane would have been the better choice for production instead of the V9/V13/D.VI biplanes. All these types used the same engine, but the V17 was the fastest fighter at the First Fighter Competition, and was described by pilots as "exceptionally maneuverable". It also had an excellent climb rate and out-performed the D.VI biplane in all respects due to its lower drag.

Despite its partially obstructed downward field of view for the pilot, the V17 was available months before the V28, which led to the production E.V. The V17's configuration also had slightly less drag than the V28 due to the V28's wing struts.

The Fokker D.VI was never intended for a major production order due to the better high-altitude performance of the Mercedes-powered D.VII. Apparently production of the D.VI was partly to absorb production of the Ur.II rotary engine. Given that substantial production was not intended, it seems even more compelling that the higher performance V17 should have been produced in preference to its biplane sibling. Production of the V17 would have provided the opportunity to evaluate the higher-performance monoplane in operational service and would have offered a faster interceptor to the *Kests* that mostly received the D.VI.

Moreover, while the D.VI choice was low risk; it was also low reward. *Idflieg* knew the D.VI was of limited use because it was basically a biplane version of the Dr.I so was a slightly faster version of that aircraft, and the D.VI was not expected to fill a crucial role in German aviation. If the V17 had been chosen instead there is a possibility it would have failed, but the failure would have been in the secondary role it was ordered to fill. However, had it succeeded, it might also have been able to fulfill a frontline role supplementing the D.VII.

Finally, the V17 and D.VI both used the 9-cylinder Ur.II rotary of 120 hp. These aircraft needed more power for more speed and even better climb rate, and the 11-cylinder Ur.III rotary rated at 145/160 hp passed its type approval test in April 1918. However, this engine was not installed until the week of 8 October in a batch of D.VIII fighters, possibly because preparing for mass-production of the engine had taken the usual timespan of four to five months. It simply arrived too late in the war to be able to show off its true potential in the D.VIII.

Fokker Fighter Frontbestand																						
Type	1915						1916						1917						1918			
	28 Feb	30 Apr	30 Jun	31 Aug	31 Oct	31 Dec	28 Feb	30 Apr	30 Jun	31 Aug	31 Oct	31 Dec	28 Feb	30 Apr	30 Jun	31 Aug	31 Oct	31 Dec	28 Feb	30 Apr	30 Jun	31 Aug
E.I			4	17	23	26	12	22	3	1												
E.II				2	8	14	20	13	4	5	2	1					2					
E.III				3	23	40	67	110	101	64	28	11	2	1	1		8	7				
E.IV					1	6	6	28	29	25	16	2										
E.V																						80
Dr.I																2	17	35	143	171	118	65
D.I										10	74	4	5	5	6	2	1					
D.II										10	49	68	49	33	26	10			6			
D.III										7	6	34	7			1		4		11	11	10
D.IV										2		2					4					
D.V													3	6	4	4	4	12	25	19	6	3
D.VI																					21	27
D.VII																				19	407	828
Total			4	22	55	86	105	173	137	137	124	112	66	45	37	19	36	58	174	220	563	1013

Above: The *Frontbestand* inventory of all Fokker fighters at the front shows the early dominance of the Fokker fighters, followed by their decline from mid-1916 until early 1918. The Dr.I triplane gave Fokker a small boost, but by the end of the war the Fokker D.VII was by far the most numerous German fighter and the D.VIII had started to arrive at the front. The few Fokker A and B-types are not shown.

Above: Captured Fokker D.VII of *Jasta* 18 on display after removal of its engine. The lack of fabric covering shows details of the aircraft's welded steel tube structure and thick airfoil wooden wings.

Fokker Aircraft Type Summary

Fokker Designation	Military Designation	Notes
Spinne	None	
M.1		
M.2		
M.3		
M.4		
M.5K	A.III & E.I	
M.5L	A.II	
M.6		
M.7	B.I	
M.8	A.I	
M.9		
M.10	B.I/II	
M.11		Not built
M.12		Not built
M.13		Not built
M.14	E.II & E.III	
M.15	E.IV	
M.16		
M.17	D.II	
M.18	D.I	
M.19	D.III	
M.20		
M.21	D.IV	
M.22	D.V	
D VI		
V.1		
V.2		
V.3		
V.4		
V.5		Dr.I prototype
V.6		
V.7		
V.8		
V.9		D.VI prototype
V.10		

Fokker Designation	Military Designation	Notes
V.11		D.VII prototype
V.12		
V.13		D.VI prototype
V.14		
V.15		
V.16		
V.17		
V.18		D.VII prototype
V.19		
V.20		
V.21		
V.22		
V.23		
V.24		
V.25		
V.26		
V.27		
V.28		E.V/D.VIII prototype
V.29		
V.30		
V.31		
V.32		
V.33		
V.34		
V.35		
V.36		
V.37		
V.38		C.I prototype
V.39		
V.40		
V.41		
V.42		
V.43		
V.44		
V.45		

Fokker V9

Above: The Fokker V9, Works Number 1831, photographed at the First Fighter Competition. (Peter M. Grosz Collection/STDB)

The order to commence work on the Fokker V9 (Works Number 1831) experimental biplane was placed to the Fokker experimental workshop on 24 August 1917, just as the first Fokker Triplanes (then still carrying the F.I designation) were about to reach the front.

The fuselage design of the V9 mirrored that of the Triplane; the fuselages of both aircraft were almost identical and they were powered by the same engine. Had both aircraft been developed at the same time, it would not be wrong to say that the V9 was the cantilever biplane fraternal twin of the F.I/Dr.I.

As it was, the V9 was completed in October 1917, when production of the Dr.I was getting into full swing. It participated in the First Fighter Competition at Adlershof in January 1918, still powered by the 110 hp Oberursel Ur.II rotary engine.

The biplane configuration reduced drag compared to the Dr.I, so was a bit faster. Its performance, particularly at high altitude, was limited by use of the low-power rotary engine, and the fact that the low-compression Oberursel engine lost power with altitude faster than the inline engines then available in Germany. More advanced rotary engines, such as the Siemens-Halske Sh.III, which retained much of its power at increasing altitudes, were not yet cleared to enter mass-production. The aircraft's low aspect-ratio wings, while very sturdy, also increased induced drag, further limiting performance at altitude.

After some further development of the wing and strut design, it was later entered into production under the military designation Fokker D.VI.

Fokker V9 Specifications

Engine:	110 hp Oberursel Ur.II	
Wing:	Span	7.70 m
	Area	16.5 m^2
General:	Length	5.90 m
	Height	2.80 m
	Empty Weight	381 kg
	Loaded Weight	572 kg
Climb:	1000m	2.5 min
	2000m	4.7 min
	3000m	7.7 min
	4000m	11.0 min
	5000m	15.5 min

Fokker V 9

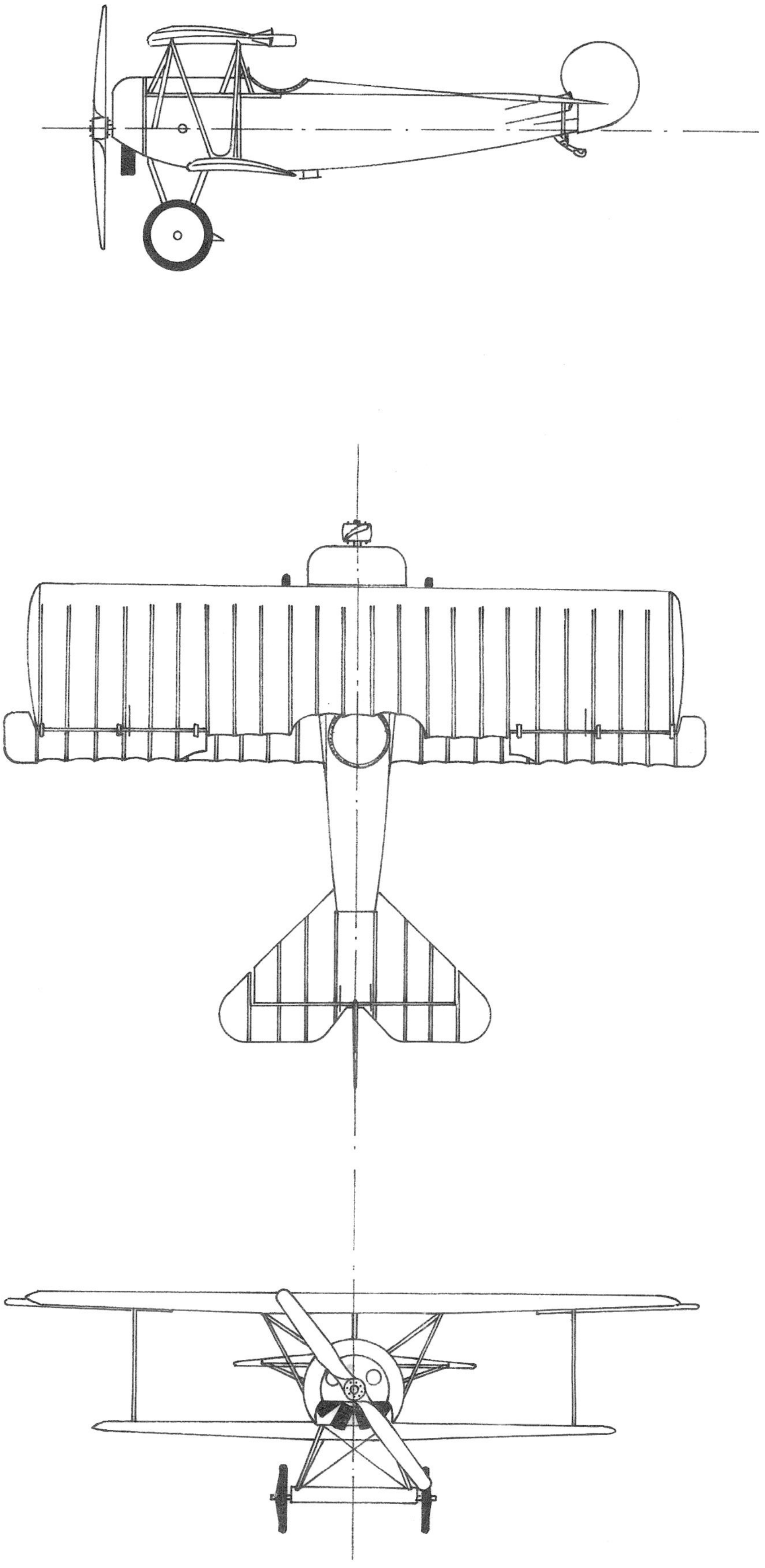

Above: This photo of the Fokker V9 shows its heritage as a derivative of the Dr.I triplane. The fuselage and tail are essentially the same with the minor differences required by its biplane configuration.

Above: The Fokker V9 prototype at Schwerin; the worker and tail support trestle have been retouched in a rough manner. The cabane and interplane strut layout as well as the wing design foreshadows that of the Fokker D.VII.

Above & Below: Front and rear views of the Fokker V9 shows Fokker's attempt at streamlining.

Fokker V10

The V10 was a design for a single-seat triplane fighter on floats powered by a 160 hp Mercedes D.III. The aircraft was commissioned from Fokker on 18 September 1917. Given w/n 1852, the aircraft was cancelled before completion in October because Fokker assessed the results of all triplanes with in-line engines as negative. The Navy wanted a land combat single-seater suitable for use in Flanders; the requirement was also referred to as the Channel task. Since reliable rotary engines of the 160 hp class were not yet available, Rolshoven considered the V6 triplane with Daimler D.III that was currently being tested, but the cooling problems and the unfavorable arrangement of the lower wing would have to be eliminated. Wing area was 22.5 or 21.4 m^2; in comparison the lighter Dr.I had a wing area of 18.66 m^2

The operational success of the Brandenberg W12 two-seat floatplane fighter eliminated further interest in single-seat floatplane fighters.

Fokker V11

Keen to evaluate the potential of the new cantilever wing design, Fokker ordered another biplane fighter from his workshop on 20 September 1917. This particular aircraft was assigned the works number 1883, and was given the Fokker designation V11. The basic design of its airframe very closely resembled that of the V9, being enlarged to make up for the additional size, weight, and power of the inline engine. Available photos taken during construction of V11show that it was, initially at least, powered by a Mercedes D.III engine, which shows clear signs of previous use. This invites suggestion that this engine may have been second-handed from another Fokker aircraft, possibly a D.IV which had been returned to the factory, or maybe even from the V2, V3, V6, or V8 prototypes.

The rear fuselages of both the V9 and the initial V11 very closely resembled that of the Fokker Triplane, which seems a reasonable design concept considering the initial successes demonstrated by the two Fokker F.Is at the front.

Unlike the V9, the V11 sported no undercarriage axle wing, indicating this component may have been considered better-suited for the rotary-engined fighters.

The first performance and weight figures for this aircraft were recorded at Schwerin as early as mid-October 1917, indicating that the aircraft was completed roughly three to four weeks after the experimental workshop received their order.

Conventional wisdom has it that the aircraft was shipped to Adlershof to take part in the First Fighter Competition on 18. January 1918, a date confirmed by Fokker records.

Fokker V11-II Specifications

Engine:	160 hp Mercedes D.III	
Wing:	Span	8.72 m
	Area	24.0 m^2
General:	Length	6.90 m
	Height	3.05 m
	Empty Weight	592 kg
	Loaded Weight	821 kg
Climb:	1000m	3 min
	2000m	6.5 min
	3000m	11 min
	4000m	17.3 min
	5000m	22.5 min

At the competition it was reportedly test-flown by Manfred *Freiherr* von Richthofen (and others), who universally praised its performance but criticised its lack of directional stability, especially in a dive, as well as its tendency to swing in flight.

In order to remedy this deficiency, Fokker supposedly rushed over his most experienced welders from Schwerin in order to lengthen the fuselage "over the weekend", and voilà, the modifications turned V11 into the clear winner of the competition.

While this makes for an amazing narrative, some indications make this story somewhat hard to believe. To begin with, it would imply that the pretty obvious defects mentioned had escaped the attention of Fokker and his team of test pilots at Schwerin over a period of two months.

Below: The Fokker V11, works number 1883, participated in the First Fighter Competition powered by a 160 hp Mercedes D.III engine. It is seen here in its original form with short fuselage and no fixed fin, clear indication of its Triplane heritage. (Peter M. Grosz Collection/STDB)

Fokker V 11

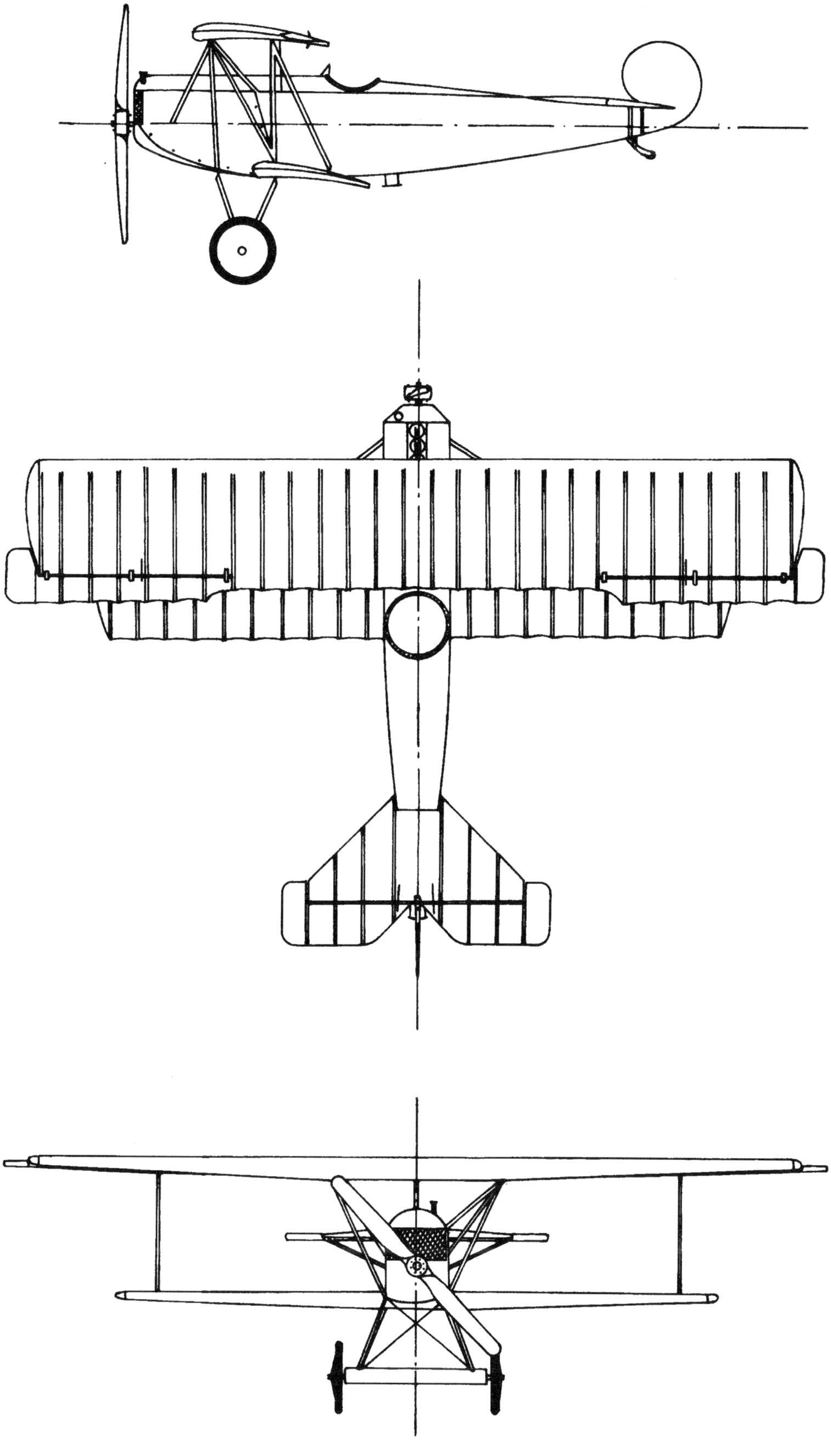

Above & Below: Fokker V11 in its original form. The short fuselage without fixed fin made the aircraft directionally unstable and challenging to fly; it was certainly not suitable for operational use.

Above & Below: Fokker V11 in its original form. There is slight dihedral on the upper wing and the undercarriage axle does not have an airfoil. The nose radiator was simple and effective. The wing design was excellent both structurally and aerodynamically, being strong, relatively simple, and offering exceptional stall characteristics. (above Peter M. Grosz Collection/STDB)

Above: The Fokker V11, work number 1883, in revised form with longer fuselage that imparted greater stability. (Peter M. Grosz Collection/STDB)

Facing Page: Fokker V11 under construction in the factory.

This seems somewhat unlikely, and even more importantly, all available photos of V11 in its initial shape were taken at Schwerin. No known photo taken at Adlershof shows the aircraft in its initial guise.

Furthermore, when comparing photos of the freshly-completed V11-I to the modified V11-II, which was photographed at Adlershof, it becomes obvious that the modifications did not just comprise of lengthening and deepening the rear fuselage. The wing area was reduced by integrating the ailerons into the wing, adding a cut-out to the wing centre-section, and possibly reducing the chord of the wings. A larger-capacity radiator was fitted, and a 170 hp Mercedes D.IIIa engine now powered the aircraft – the upright air pump ahead of the first cylinder is now clearly visible. The familiar "streaked" camouflage scheme had been applied, but no works number is visible.

A look at a graph documenting the wing loads of Fokker aircraft shows that a wing load of approximately 34,2 kg/m^2 was recorded for V11 in mid-October 1917. By early January, before the aircraft was shipped to Adlershof, a wing loading of around 41,29 kg/m^2 was recorded for V11. This clearly indicates that by then, the lengthened fuselage and reduced wing area (now 20.2 m^2) had been applied to the aircraft, resulting in V11-II. Interestingly, the wing area of V11-II was now exactly that of the D.VII production aircraft.

Taking all this into account, the aircraft appeared markedly different from its first appearance, and to this day it is not entirely clear if the photos show a radically-rebuilt airframe or two completely different planes. Certainly, Fokker records only mention the works number 1883 for V11, implying that it was indeed just one aircraft.

On the other hand, the sheer extent of modifications, which literally cover almost the entire airframe, suggests the existence of two aircraft under the V11 designation.

After Fokker received his initial production order for the D.VII, V11 was included into this batch of 300 aircraft where it was assigned the military number 227/18. This way, Fokker obtained payment from the authorities for an experimental aircraft which he had originally financed out of his own pocket. All indications are that V11 was destroyed during the type-test procedure at Adlershof in February 1918, where load-tests were carried out on the airframe, proving it to exceed all specified load factors.

The question whether the V11 or the V18 was the true prototype of the D.VII is impossible to answer. While the V11-II was picked as the winner of the First Fighter Competition, the production D.VII inherited features from both of these aircraft, and a few entirely new details as well. In a way, the V11 and V18 were two parallel paths that eventually merged into the Fokker D.VII by late February 1918.

1064.

1065.

Left & Below: Wing cellule of the Fokker V11 prototype being stress-tested in the Fokker factory.

This Page: Components of the Fokker V11 wing cellule being stress-tested in the Fokker factory.

Fokker V12

Above: A factory photo of Fokker V12, Works Number 1980, a D.VI prototype and a sister ship to experimental Triplane V.7R, WN 1981. Both were shipped to Hungary on 3 January 1918 and fitted with identical cowlings, Steyr-LeRhônes, and propellers. Fouteen V12s were originally ordered as D.VIs, and seven were delivered and fitted with A-H Steyr-LeRhônes, props, and Schwarzlose machine guns. The other five were converted to D.VII orders. The D.VIs continued to fly until the last was lost in 1926.

Fokker V12 Specifications

Engine:	110 hp Oberursel Ur.II	
Wing:	Span	7.68 m
	Area	24.0 m^2
General:	Length	7.50 m
	Height	3.05 m
	Empty Weight	592 kg
	Loaded Weight	821 kg
Climb:	1000m	3 min
	2000m	6.5 min
	3000m	11 min
	4000m	17.3 min
	5000m	22.5 min

The Fokker V12 (Works Number 1980) was delivered without engine and armament to Austria-Hungary on 3 January 1918 and was equipped at MAG with the 150 hp Steyer-Le Rhône. The probable military number of the V12 in Austro-Hungarian service was 90.04. In June, based on the good flight results of the V12, the *LFT* ordered 14 series D.VIs from Fokker in Schwerin, seven of which were delivered. Works Number 1980 participated as "MAG-Fokker 150 hp Le Rhône biplane" in the Austro-Hungarian fighter competition in Aspern in July. In the accounts of the disbanded Austro-Hungarian air force after the end of the war, Works Number 1980 was designated V12. When it was commissioned on 22 October 1917, it was still running as a V13 with a 160 hp Siemens Sh.III engine. At that time, on 29 October 1917, another machine V12 [Works Number 1982] was ordered as a biplane with 110 hp Ur.II. In a Fokker compilation of May 1919, Works Number 1982 was then designated V13, which was sent to Budapest

Above: Fokker V12 in Hungary after being fitted with a 150 hp Steyer-Le Rhône and new propeller. The works number 1980 is barely legible on the engine cowling. At this point, no armament has been installed. (Peter M. Grosz Collection/STDB)

together with the V7 [Works Number 1981] on 3 January 1918. It is unclear whether the contradictory information is only a spelling error in the little documentation available or was as convoluted as described, whereby no systematic designations of prototypes seem to be recognizable.

Fokker V13

Above: Fokker V13-II, Fokker Works number 2054, the second of two V13 prototypes, participated in the First Fighter Competition powered by a 160 hp Siemens-Halske Sh.III. The V13-I, works number 1980, was powered by the 145 hp Oberursel UR.III. The Fokker V13 prototypes featured the N-strut design used by the production Fokker D.VI. (Peter M. Grosz Collection/STDB)

At least three different biplanes carried the V13 designation, possibly only temporarily. Probably some of these are errors in the surviving Fokker documents. Works Number 1980 was sent together with the V7 (Works Number 1981) to Budapest for MAG on 3 January 1918, where both received the Steyr-built Le Rhône. In the 1919/1920 period, Works Number 1980 was also referred to as the V12. Works Numbers 1983 and 2054 flew as the V13-I and V13-II at the First Fighter Competition at Adlerhof in January 1918, the latter with a pre-series 160 hp Siemens & Halske Sh.III engine. Works Number 1983 was originally ordered on 2 November 1917 as the V14 with 160 hp Le Rhône. The order for Works Number 2054 as the V13 biplane was placed on 26 November 1917. On the first day of the Adlershof competition (21 January 1918), Anthony Fokker impressed the *Idflieg* representatives with a dramatic aerobatic flight in the V13-I. When the official climb measurements by *Idflieg* began on January 25, the V13-I went out of control while the pilot was adjusting its prototype Ur.III engine and rammed the Fokker V18, but the damage to its lower wing was quickly repaired. On January 26, 1918, Fokker pilot Schützenmeister reached an altitude of 6000 meters on the V13-I in 22.7 minutes. However, the engines of the two V13s in the competition, which had not yet been officially approved for production, caused many an aborted flight. In terms of construction, the V13 was a V9 with a slightly extended fuselage.

Fokker V13-I Specifications

Engine:	145 hp Oberursel Ur.III	
Wing:	Span	6.75 m
	Area	17.4 m^2
General:	Length	6.35 m
	Height	2.82 m
	Empty Weight	472 kg
	Loaded Weight	644 kg
Climb:	2000m	4.5 min
	3000m	7.7 min
	4000m	11.0 min
	5000m	16.2 min

Fokker V13-II Specifications (as V13-I except)

Engine:	160 hp Oberursel Ur.III	
	Height	3.20 m
	Loaded Weight	669 kg

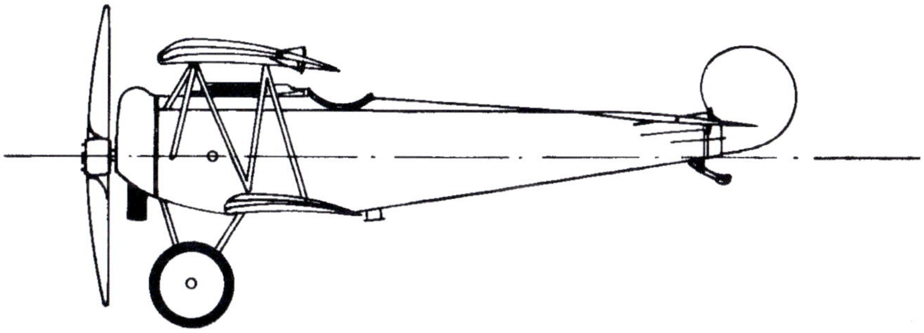

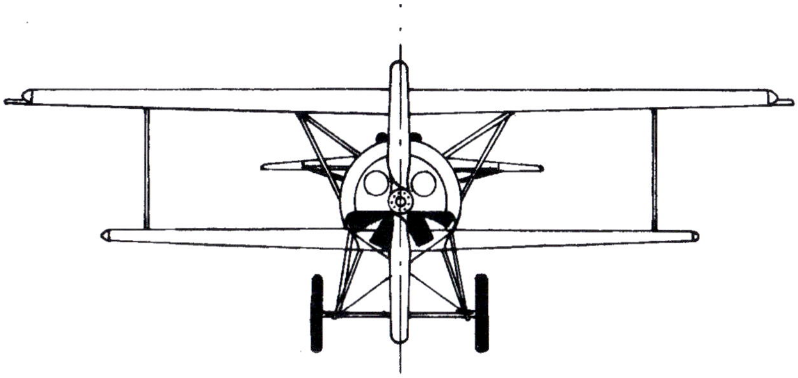

Fokker V14

Works Number 1983 was ordered on 2 November 1917 with the designation V14.

V14 was a biplane with UR.III rotary and participated as a V13 in the first fighter competition at Adlershof.

Fokker V15

There is no data available on the V15.

Fokker V16

The design of the V16 was similar to the V7 triplane with the middle wings omitted to explore the flight characteristics; the upper and lower wings were enlarged.

V16, w/n 2085, was ordered on 3 December 1917 with a 110 hp Oberursel UR.II rotary engine. The wings were cantilever and their surface area was specified as 16.6 m^2 in the documentation, slightly smaller than the 18.66 m^2 of the production Dr.I.

The purpose of V16 was to investigate the aerodynamic effect of a larger wing gap. The result was deterioration and flight testing was stopped.

Fokker V17 Specifications

Engine:	110 hp Oberursel UR.II	
Wing:	Span	8.70 m
	Area	16.8 m^2
General:	Length	5.80 m
	Height	2.95 m
Max speed:		200 km/h
Climb:	2000 m	4.5 min.
	3000 m	7.5 min.
	4000 m	11.7 min.
	5000 m	17.7 min.

Fokker V17

Fokker's first cantilever monoplane fighter, the V17, Works Number 2147, was the final rotary-powered Fokker entry in the First Fighter Competition. The construction order of 28 December 1917, issued only three weeks earlier, lists the 110 hp Le Rhône engine as the powerplant. The idea for the development of cantilever monoplanes certainly came from Fokker's experience flying the breakthrough all-metal J2 monoplane in December 1916 and the two-seat all-metal Junkers J8 a year later on December 16, 1917.

Only three weeks after construction began, the V17 went to the first fighter competition at Adlershof on January 17, 1918, but was subsequently not pursued further. Fokker was able to dispel the pilots' initial distrust of the monoplane with his aerobatic demonstrations on the V17 at Adlershof.

The wooden wing had a box spar covered in plywood, giving it a smooth finish and reasonable torsional stiffness for effective aileron control. Powered by the same 110 hp Oberursel UR.II engine used in the Fokker Dr.I Triplane, it was reportedly the fastest fighter at the First Fighter Competition. and was described by pilots as "exceptionally maneuverable". It also had an excellent climb rate; however, its wing obscured the pilot's field of view downward, particularly important during takeoff and landing, and it was officially rejected for that reason. A more important reason may have been that monoplanes, especially cantilever monoplanes, were viewed as structurally suspect. On 17 October 1918, after repairs, the aircraft was delivered to ZAK at Johannisthal.

Fokker V17 Specifications

Engine:	110 hp Oberursel UR.II	
Wing:	Span	8.375 m
	Area	11.6 m^2
General:	Length	5.80 m
	Height	2.80 m
	Empty Weight	356 kg
	Loaded Weight	535 kg
Climb:	1000 m	3.2 min.
	2000 m	6 min.
	3000 m	9 min.
	4000 m	13 min.
	5000 m	19 min.

Above: The Fokker V17 participated in the First Fighter Competition powered by a 110 hp Oberursel UR.II. Officially rejected due to inadequate field of view downward, the preference for sturdy biplanes may also have influenced the decision. The Fokker D.VI biplane, powered by the same engine as the V17, was selected for limited production after competing with the V17 at the First Fighter Competition despite the V17 offering better speed and climb rate. It is interesting to speculate on the combat success of the V17 had it been produced in quantity for operational service, which could have happened five or six months prior to the appearance of the later Fokker E.V that was still powered by the same low-power engine. By its rejection of the V17, *Idflieg* may have missed an excellent opportunity. Inability to bring the 11-cylinder, 145 hp UR.III rotary into timely production was also a major impediment to all of Fokker's 1918 rotary fighter designs, which needed more power. The UR.II had only ten more horsepower than the U.I of 1915! The first acceptance of a production fighter with the 145 hp UR.III, a Fokker D.VIII, was October 8, 1918, a mere month before the Armistice.

Fokker V 17

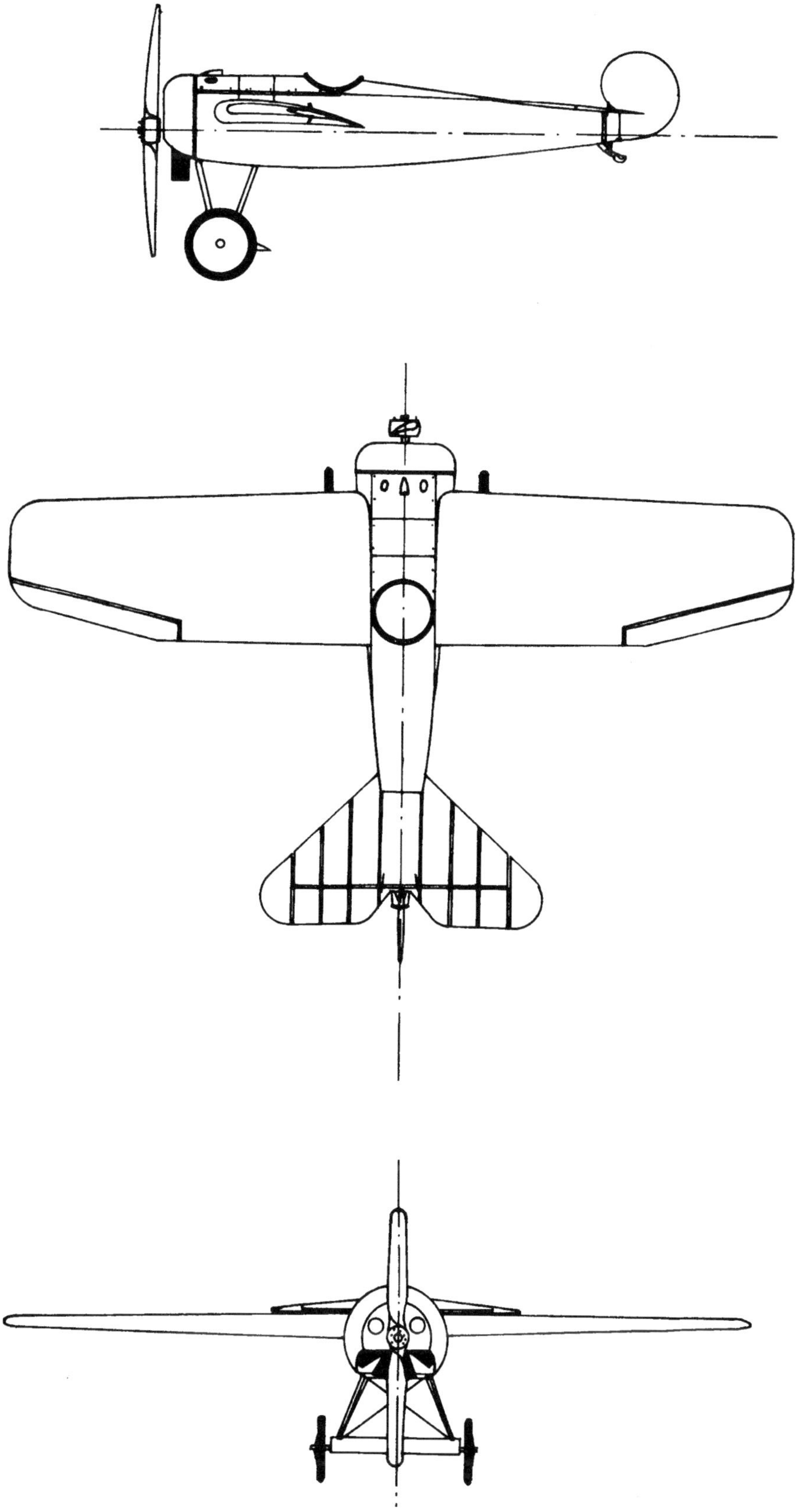

Above & Below: The Fokker V17, work number 2147, powered by a 110 hp Oberursel UR.II participated in the First Fighter Competition. Despite its low power it was the fastest fighter at the competition. It also had a good climb rate and was very maneuverable. The V17 was the first Fokker cantilever monoplane design. The wood wings were built around a box spar covered with plywood, providing strength and torsional stiffness for effective aileron control. The wing obscured the pilot's downward view and monoplanes were structurally suspect, so the V17 remained a prototype.

Above & Below: The Fokker V17, work number 2147, powered by a 110 hp Oberursel UR.II participated in the First Fighter Competition. Here it is parked next to an AEG G.V to contrast the difference in size.

Fokker V18

The Fokker V18 experimental aircraft, given the Fokker works number 2116, was closely related to the V11 and must be seen as the second prototype ancestor to the aircraft that eventually entered serial production under the military designation Fokker D.VII.

From the start, this machine had a taller rear fuselage that already closely resembled that of the later D.VII. Importantly for its flying qualities, it had much enlarged vertical tail surfaces for better stability. The leading edge of the horizontal tailplane was curved, resembling the shape of the component on the early Fokker F.Is. As was the case on the Triplane, the curved leading edge of the horizontal tail surfaces was not adopted for production aircraft.

The construction order for the Fokker V18 was dated 12 December 1917; work on the aircraft proceeded quickly. It was sent to Adlershof on 17 January 1918, actually a day before the V11 was, in order to participate in the First Fighter Competition. Once again, Fokker records document the aircraft had a 160 hp Mercedes D.III, while photos prove it was in fact powered by a D.IIIa.

Without a doubt, Fokker wanted to assure that his latest designs were powered by the better-performing engine during the competition.

On 25 January 1918, V18 was damaged as a result of an accident. As the result of an Oberursel maintenance man making a mistake while adjusting the engine of Fokker V13, which was parked nearby, the rotary-engined prototype made a sudden hop that was halted by the tip of the lower left wing of V18. However, the damage was quickly repaired, and testing of V18 resumed on 27 January.

Like V11, the aircraft was missing the axle wing that was a standard component on the Fokker Dr. I, as well as on all of Fokker´s 1918 production aircraft.

After Fokker received his initial production order for the D.VII, the V18 prototype was absorbed into this batch of 300 aircraft, where it was assigned the military number 228/18. However, it is unknown if it was ever fully brought up to D.VII production standards, or if this inclusion into the first batch was merely intended to assure that Fokker would receive payment for his prototype investment.

Fokker V18 Specifications

Engine:	170 hp Mercedes D.IIIa	
Wing:	Span	8.72 m
	Area	23.1 m^2
General:	Length	7.03 m
	Height	2.9 m
	Empty Weight	634 kg
	Loaded Weight	860 kg
Climb:	2000m	5.3 min
	3000m	10.5 min
	4000m	17.3 min
	5000m	26.0 min

Above: The Fokker V18 was based on the same formula as the V11 but had larger vertical tail surfaces for improved stability.

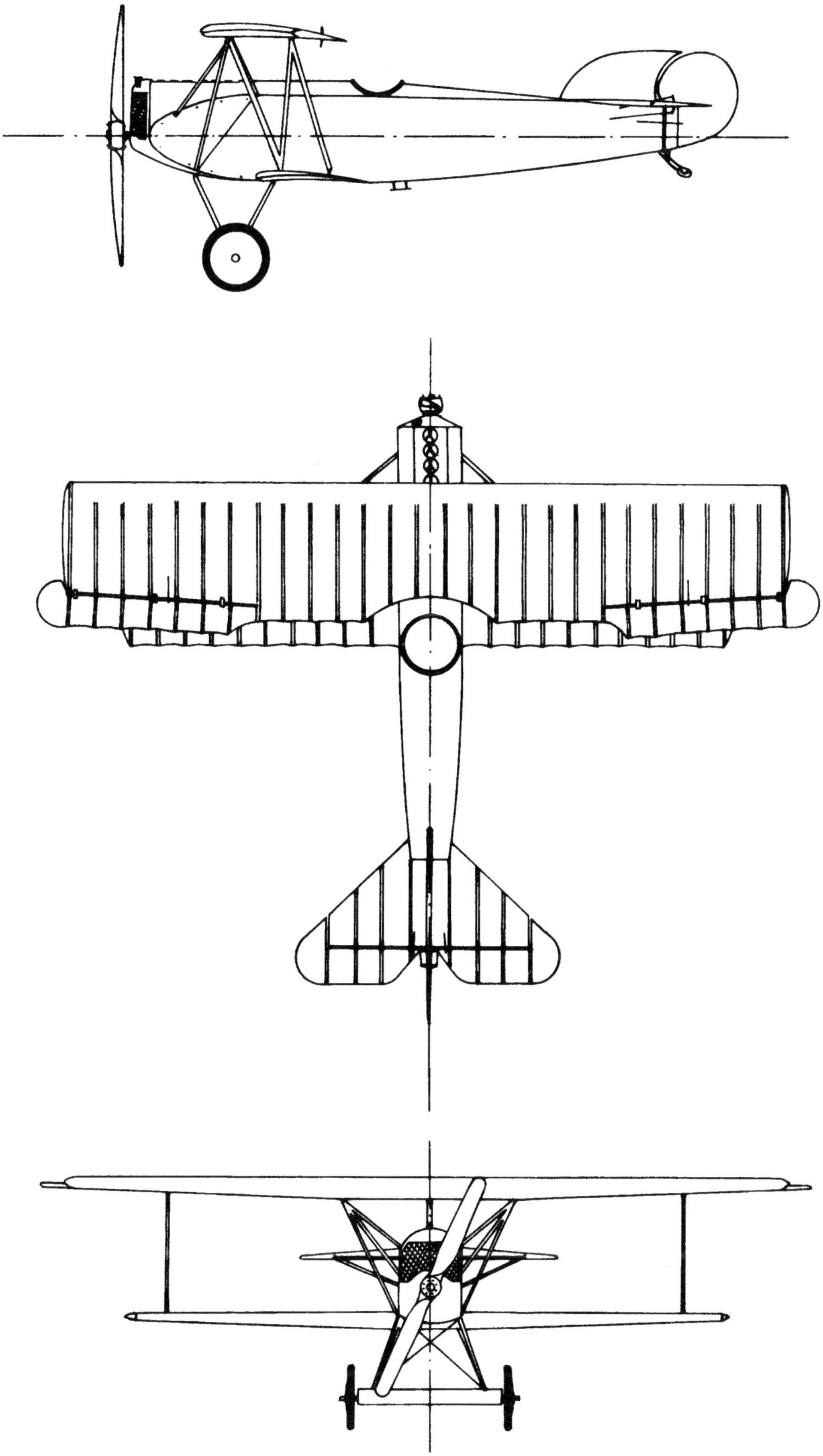

Above: The Fokker V18 was based on the same formula as the V11 but had larger vertical tail surfaces for improved stability.

Above: Closeup of the exhaust side of Fokker V18 showing noticeable stains on the rear cabane strut. Of interest is the cut in leading edge of the metal side panel which enabled swivelling the component back into place after engine maintenance.

Above: Fokker V18, works number 2116, at the First Fighter Competition at Adlershof in late January/early February 1918. The curved leading edge of the horizontal tailplane is in evidence. This was of a very similar shape to that seen on the Fokker F.I Triplanes. The somewhat ungainly tail fin appears to have been retro-fitted, and its shape loosely resembles a similar component seen on the earlier Fokker D.IV. Also apparent in this view is the comparatively large cut-out to the upper wing centre-section.

Above: In this front view, V18 closely resembled its sister aircraft V11. The light-colored tip of the lower left wing indicates this photo, as well as the other frontal view, were taken after the damage suffered on 25 January had been repaired.

Fokker V19

According to the Fokker type list, the V19 was a single-seat monoplane with 110 hp UR.II, about which no further information is given.

Fokker V20

The V20, Works Number 2219, was a mid-wing cantilever monoplane with Mercedes D.III engine for better high-altitude performance. Fokker reports that he ordered the V20 as a variant of the V17 with the more powerful in-line engine in his experimental department only in the course of the first Adlershof competition and that it was built within only one week under the direction of Reinhold Platz in Schwerin and delivered to Berlin, where Fokker demonstrated it.

According to Platz's recollections, Fokker called him from Berlin on a Saturday (26 January 1918) and on the following Saturday (2 February 1918) the plane was ready and was

Fokker V20 Specifications

Engine:	160 hp Mercedes D.III	
Wing:	Span	9.29 m
	Area	15.2 m^2
General:	Length	7.135 m
	Height	3.0 m
	Empty Weight	620 kg
	Loaded Weight	821 kg
Climb:	2000 m	6.7 min.
	3000 m	11.7 min.
	4000 m	18.0 min.
	5000 m	26.7 min.

Above & Below: Fokker V20 monoplane prototype, work number 2219, was the inline-powered counterpart to the V17 rotary-powered prototype. The V20 was designed, built, and flown in the incredibly short time of 6½ days during the First Fighter Competition.

Fokker V 20

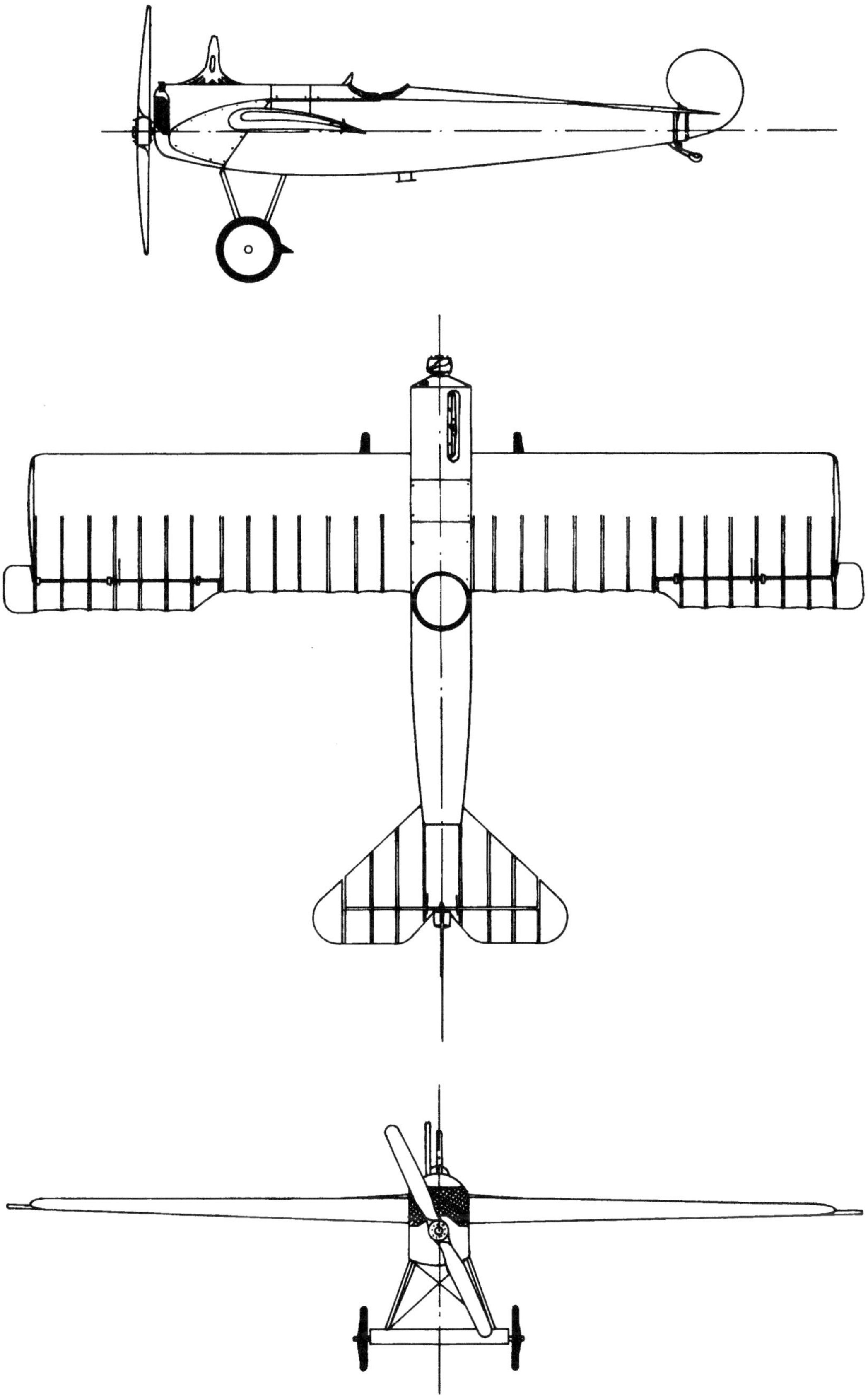

Above & Below: These photos of the front and rear of the inline-engined Fokker V20 show how streamlined the design was for 1918.

Above: Like the V17, the wing of the V20 obscured the pilot's field of view downward, and monoplanes, especially cantilever monoplanes, were viewed as structurally suspect, so it was not produced in quantity. Instead, its biplane counterpart the V18 served as one of the prototypes of the Fokker D.VII.

flown in by pilot Weidner due to Fokker's absence. The *Idflieg* monthly report for March 1918 mentions a Fokker monoplane with an inline engine in flight testing, but states: "The aircraft in its present configuration cannot be used as a front-line aircraft, as the visibility is too poor. For the cantilevered wooden wing with one spar, Fokker documents also state 16.0 m^2 wing area.

Fokker V21

For many years, the true identity of the Fokker V21 had researchers puzzled. It was known that this aircraft had been assigned the Fokker Works Number 2310, but since a rudder showing just this works number was seen on a Mercedes-powered mid-wing monoplane, it was determined that this must have been the V21. Thanks to the research of the much-missed historian Alex Imrie we know that this mid-wing monoplane was, in fact, Fokker V23, Works Number 2443. When said photograph was taken, V23 mounted a "comma"-shaped rudder which was originally fitted to the real V21, and this component was marked with the Fokker Works Number 2310.

The true Fokker V21 was, in fact, the Fokker D.VII production prototype, and it is termed exactly that in official Fokker records. Unfortunately, the few available photographs of V 21 only show it being fitted with the final tail-plane, featuring the tall rudder complete with the fin as seen on the production Fokker D.VIIs.

In these photos the aircraft is painted in the overall "streaked" scheme known so well from the Fokker Dr.I, which was still in full production at Schwerin when the V21 first appeared in February 1918. Records documenting wing data and engine performance mention the D.VII for the first time during this month, and in this context it is possible to determine that V21 was indeed the D.VII "pattern aircraft".

It is noteworthy that the Iron Cross markings were also applied in an identical way to that seen on the Dr.I, being marked on a rectangular white background, a style of application that had become obsolete by an *Idflieg* instruction dating back as far as October 1916!

Fokker V21 Specifications

Engine:	160 hp Mercedes D.III	
Wing:	Span	8.90 m
	Area	20.2 m^2
General:	Length	6.945 m
	Height	2.945 m
	Empty Weight	652 kg
	Loaded Weight	870 kg
Climb:	2000m	7.0 min
	3000m	12.0 min
	4000m	19.5 min
	5000m	27.0 min

Above: The Fokker V21 was the production prototype of the Fokker D.VII.

Fokker V 21

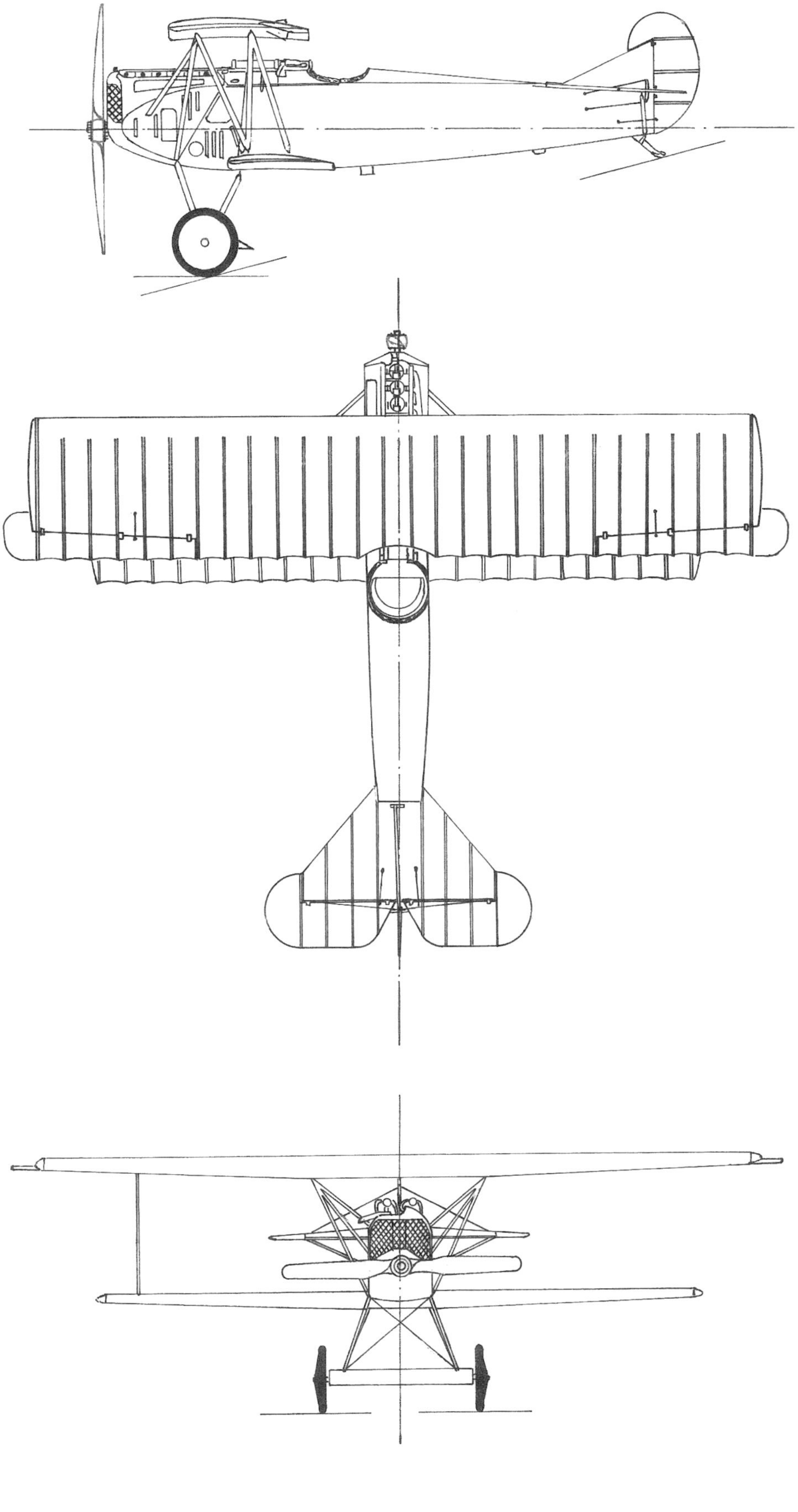

Above: The Fokker V21 was a variant of the Fokker D.VII. Old-style national insignia were applied.

The fact that the initial small rudder, which was apparently first fitted to V21, was then transplanted to V23 proves that even by early February the final D.VII production configuration had not yet been determined. Last-minute modifications were still being carried out then.

Streaky fabric covering aside, the details seen on V21 match those of the first production D.VIIs, including the wing layout and the type of radiator fitted to the airframe. Some very minor details differed, but on the whole it can be seen as the true "pre-production" Fokker D.VII. Unsurprisingly, the overall dimensions recorded for V21 match those of the D.VII.

On 22 May, V21 aka Werknummer 2310, was sent to *Flz.A.Gr.*6 (the aircraft testing department) at Adlershof, in order to participate in the upcoming Second Fighter Competition.

We hear about it one more time on 17 October, when it was sent to the Central Acceptance Commission at Berlin-Johannisthal.

Above: Front view of the Fokker V21 was essentially identical the production Fokker D.VII except for the national insignia.

Above: The Fokker V21 was finished like the production Dr.I that was still in production when the V21 apperared.

Fokker V22

After the victory of Fokker's new D.VII fighter in the First Fighter Competition, Austrian pilots were given the opportunity to fly the first prototype. Fokker wrote to the Austro-Hungarian *Fliegerarsenal* (*Flars*) as early as 19 February 1918 that he could deliver 40 D.VIIs a month to Austria if he could get the necessary quota of steel tubes. In four weeks, he wanted to have two samples of the D.VII ready for the LFT. The first was the V22, powered by the 200 hp (19000 series) Austro-Daimler engine and a four-bladed Jaray propeller and steel tube fuselage.

The V22 carried the Fokker Works number 2342, and this reveals that it was "sandwiched" into the early Fokker D.VII fighters, between 256/18 (w/n 2341) and 257/18 (w/n 2343). As such, it was a series-production airframe that was adapted for installation of the Austrian engine and armament. This is an important fact that has to be kept in mind when reading the later objections brought up by the Austrian authorities against license manufacture of the aircraft.

A second projected prototype was to have a plywood fuselage, intended as a fallback solution in case a lack of steel tubing would prevent serial production either in Austro-Hungaria or in Schwerin as an "export" version. However, as early as March 1918, *Idflieg* officer *ObLt*. Reichelt indicated that a number of wagon-loads of steel tube could be supplied to ensure production at or for the Austro-Hungarian empire.

Fokker V22 Specifications

Engine:	200 hp Austro-Daimler (MAG)	
Wing:	Span	8.90 m
	Area	20.2 m^2
General:	Length	6.945 m
	Height	2.945 m
	Empty Weight	689 kg
	Loaded Weight	908 kg
Max Speed		210 km/h
Climb:	2000m	5.0 min
	3000m	9.0 min
	4000m	13.5 min
	5000m	19.0 min

On 2 March, Fokker requested the delivery of two 8 mm Schwarzlose M16s for the V22, which was to receive the latest version of Fokker's central control system. He had agreed to deliver the V22 readily-equipped to full military standards. Of course, this would allow for an evaluation of the type under realistic circumstances.

On 15 and 16 March 1918, Austrian *Stabsfeldwebel* Franz Kuttner flew the V22 at Schwerin-Görries. Kuttner had previously been a Fokker flight instructor. During these

Above: The V22 was a D.VII prototype with 200 hp Austro-Daimler engine and 4-bladed Jaray propeller. In an attempt to improve propeller efficiency, the blades were oriented at 70° instead of the much more common 90°. The extra power of the 200 hp Austro-Daimler engine gave it much improved performance compared to the V21.

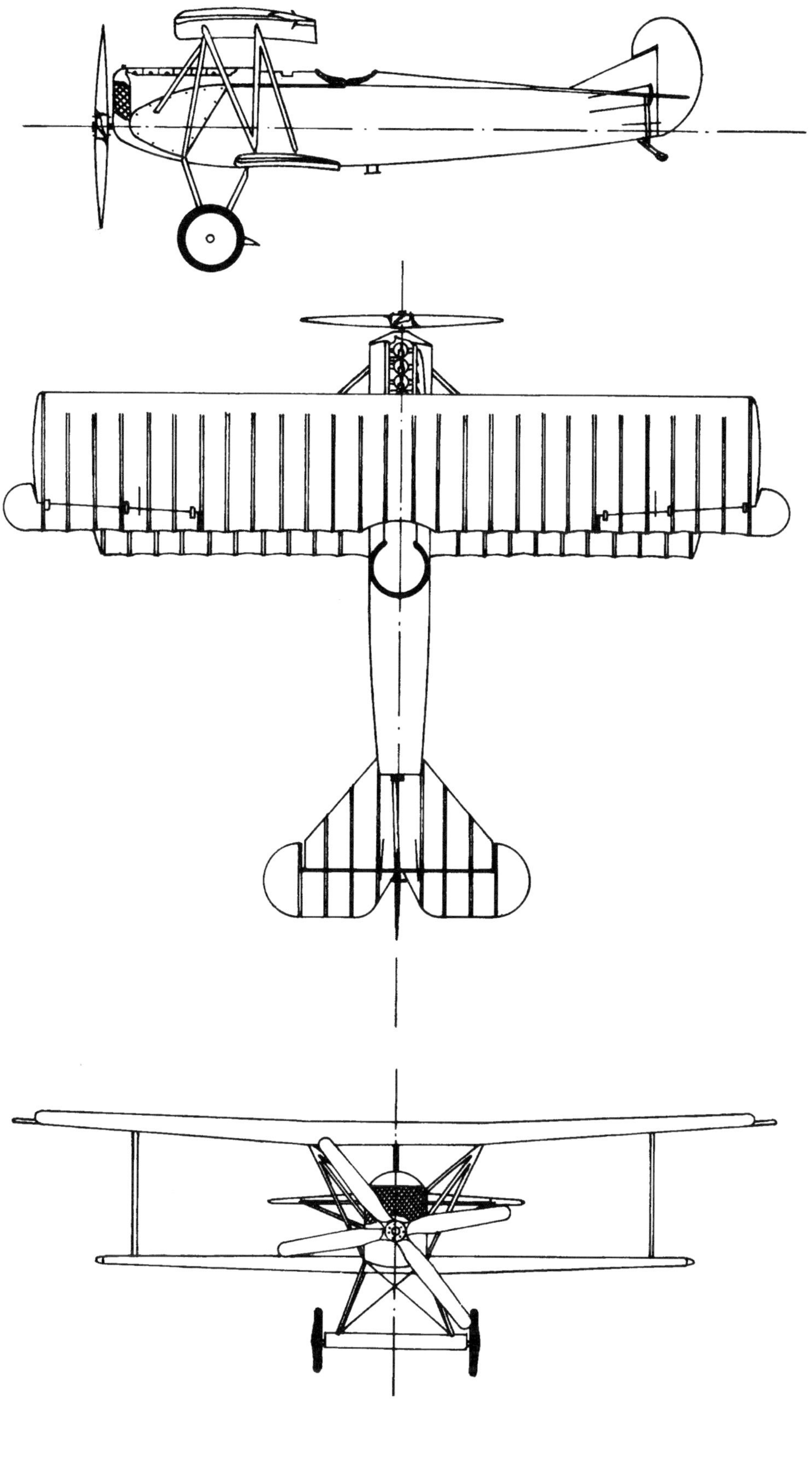

Above & Below: The V22 was a D.VII prototype with 200 hp Austro-Daimler engine and 4-bladed Jaray propeller for evaluation by the Austro-Hungarian *Fliegerarsenal*. The extra power of the 200 hp Austro-Daimler engine gave it significantly improved performance compared to the V21. The Austrian authorities were concerned about the lack of steel tube, but after Germany agreed to supply sufficient steel tube for production – and *Lt*. Mallinkrodt's demonstration of the aircraft to the LFT chief – the aircraft was chosen for Austro-Hungarian production. The all-wood Pfalz D.XII was also considered for production because Austria had plentiful supplies of wood, but the Fokker had better flying qualities.

evaluations, a climb time to 3000 meters was recorded in 8 to 9 Minutes, while 5000 meteres were reached in 17½ to 19 minutes. The aircraft was reported to have very good flight characteristics, offered remarkably good visibility as was very manouverable.

In early April, the Austrian War Ministery purchased the aircraft, which was delivered to MAG on 24 April 1918. Fokker invoiced Flars via MAG on 13 June, the price of the airframe was 30.000 Marks.

Thorough testing by the *Fliegerasenal* at Wien-Aspern followed under Austrian military serial number 90.05. *Flars* representatives were initially opposed to serial production of the aircraft, expecting equal or better performance from their own new or improved fighter designs.

An Austrian report dated 20 April listed seven points that supposedly prevented production of the type by an Austro-Hungarian manufacturer under licence. These points document an obvious ignorance of facts that can only be explained by a massive bias of the Austro-Hungarian authorities against the Fokker design. For this reason, they are repeated in full here:

"The aircraft presented cannot be considered to enter series production with us.

1) The performance should be equalled by our new single-seaters (powered) by the 225 hp Daimler and 240 hp Hiero (engines), possibly even exceeding it. (author´s note: This is a reference to the Oeffag D.III, Berg D.I, Phoenix D.III, and WKF D.I fighters)

2) By the time license manufacture is fully underway (5 months or more) our new single-seaters are already available.

3) 19000-engines (i.e. the 200 hp variant of the Austro-Daimler fitted to the V22) are no longer in production.

4) Range is less than that of our single-seaters. In that context, useful load must be reduced in order to increase the fuel capacity.

5) In spite of the lower useful load, 190 Kg versus 240 kg in our planes, the total weight is higher than that of the Berg single-seater.

6) In terms of structural strength, the Fokker will likely lag far behind our single-seaters.

7) Installation of machine guns and visibility, especially ahead at eye level, are impeccable, but should be ideally equalled by our new single-seaters."

Regarding point 1: For the most part, this remark can be interpreted as wishful thinking. The Oeffag D.III was a licence-built version of the Albatros D.III, a 1916-vintage design that can be considered completely outdated by April 1918. The Berg was structurally weak, even when powered by lighter, less powerful versions of the Austro-Daimler. Finally, the Phoenix was also a design that had reached the limits of further development.

Regarding 2: As it turned out, the only new design hinted at, the WKF D.I, would not enter frontline service before the end of the war. The other three designs, powered by the 225 hp Daimler and 240 hp Hiero engines, did see service, but were merely evolutions of old and, more or less, proven designs.

Regarding 3: This objection cheerfully ignores the fact that 225 hp Daimler engine could have been easily installed in the V22 – had an example of it been made available to Fokker. That was not the case at the time. Flars officials obviously feared that it would give the Fokker an improved performance that could not be matched even by the new domestic designs powered by the same engine.

Regarding 4: The only new design hinted at, the WKF D.I, would not enter frontline service before the end of the war. The other three designs, powered by the 225 hp Daimler and 240 hp Hiero engines, did see service, but were evolutions of old designs

Regarding 5: Available records indicate that the V22 was indeed tested with a useful load of 190 Kg. Endurance of the WKF D.I was 2 hours and 30 minutes, one hour more than that of the V22, but it weighed 1005 Kg, almost 100 KG more than the V22.

Regarding 6: By that time, the D.VII had passed all mandatory German load tests with flying colors, and since the V22 was nothing but a modified D.VII these results could be applied to the V22 as well. In spite of this, as late as 09 July, the Austrian authorities noted that the V22 was not cleared to be flown by frontline pilots since calculations concerning structural strength were not (yet) available.

The Berg D.I had suffered from structural failure of its wings during its entire frontline life, the Oeffag D.III featured a single-spar lower wing versus the V22´s double spar lower wing. All Austro-Hungarian designs had wooden fuselages that offered inferior survivability for the pilot in case of a crash compared to the steel-tube fuselage used by Fokker design.

Regarding 7: Positioning the radiator in the nose of the D.VII meant that no water pipes coming down from the upper wing obstructed the forward view for the pilot, as was the case of several of the competing designs. Mounting the machine guns on top of the fuselage alone did not ensure good forward visibility from the cockpit.

Aside from the points quoted, this report completely ignores the most important technical advance of the V22: its cantilever wing that rendered rigging wires completely obsolete. In this context it is worth mentioning that all domestic designs that are hinted at in the report (Oeffag D.III, Berg D.I, Phoenix D.III, and WKF D.I) still featured the now obsolete thin airfoil rigged wings.

V22 took part in the fighter evaluation flight in Wien-Aspern from July 9 to 23, 1918, where LFT pilots were forbidden to fly it, due to a lack of documents proving the structural integrity of the aircraft. This happened at a time when hundreds of D.VIIs were serving without any sign of structural defects in dozens of *Fliegertrupper Jagdstaffeln* on the Western Front! Therefore, Fokker's MAG-Chief Seekatz organized a presentation of the aircraft by German *Lt.* Mallinckrodt to the LFT chief. By August the examination

V22 in closeup at Schwerin.

Above: Closeup of V22 tail assembly; Fokker D.VII 241/18 is in the background.

of the Fokker D.VII, as it was now referred to, had been successfully concluded, and orders for Austro-Hungarian use could be placed, at last.

Above: V22 machine gun installation.

Above: V22 was camouflaged before it was sent to MÁG factory and designated 90.05. In this photo it has Fokker 2-bladed propeller. Fokker triplane prototype 90.03 is in the left background. (Peter M. Grosz collection/STDB)

Above: Now designated 90.05, V22 it at right. In this photo it has Fokker 2-bladed propeller. Fokker triplane prototype 90.03 is at left. (Peter M. Grosz collection/STDB)

Above & Below: V22-II with wooden fuselage & 200 hp Austro-Daimler. (Peter M. Grosz Collection/STDB)

Above & Below: V22-II with wooden fuselage & 200 hp Austro-Daimler. (Peter M. Grosz Collection/STDB)

Above & Below: V22-II with wooden fuselage & 200 hp Austro-Daimler. (Peter M. Grosz Collection/STDB)

Fokker V23

The Fokker V23, Works Number 2443, was another mid-wing monoplane design similar to the V20, which had first appeared at the First Fighter Competition at Adlershof in January 1918. Unlike the V20, it had a two-spar wooden wing, but the ailerons were inset into the wings and did not have aerodynamic balances. It was hoped that a new wing profile would improve performance over the V20, but this was not achieved to any great extent. The V23 participated in the Second Fighter Competition in May 1918, where it was shipped on 28 May 1918. In July 1918, *Idflieg* declined to place a series order for the V23, as it had for the V20. In spite of showing excellent speed and climb, the then-fashionable reservations against mid-wing monoplanes prevented approval for production. On 17 October, after the completion of unspecified repairs, it was delivered to the ZAK (Central Acceptance Commission) in Johannisthal.

In available photos of V23, the Works Number 2310 is visible on the rudder. As related in the V21 text, this was actually the Werknummer of the V21 prototype. Thus, it appears that the V 21 was initially test-flown without a tail fin. After it was modified accordingly, the surplus comma-shaped rudder was then given a second life on the V23 prototype, Works Number 2443. Obviously nobody saw the need to apply the correct works number to this component since the aircraft had no prospects of being chosen for production. At the time, no one thought of the confusion caused by this small oversight decades later.

Fokker V23 Specifications

Engine:	160 hp Mercedes D.III	
Wing:	Span	9.01 m
	Area	13.6 m^2
General:	Length	7.81 m
	Empty Weight	658 kg
	Loaded Weight	877 kg
Maximum Speed:		210 kmh
Climb:	1000m	2.5 min
	2000m	8.5 min
	3000m	12.0 min
	4000m	17.0 min
	5000m	23.0 min

Above: Officially the Fokker V23 was rejected because of the poor downward view for the pilot due to the wing location. Again, prejudice against monoplanes was probably involved also. Power was a 160 hp Mercedes D.III.

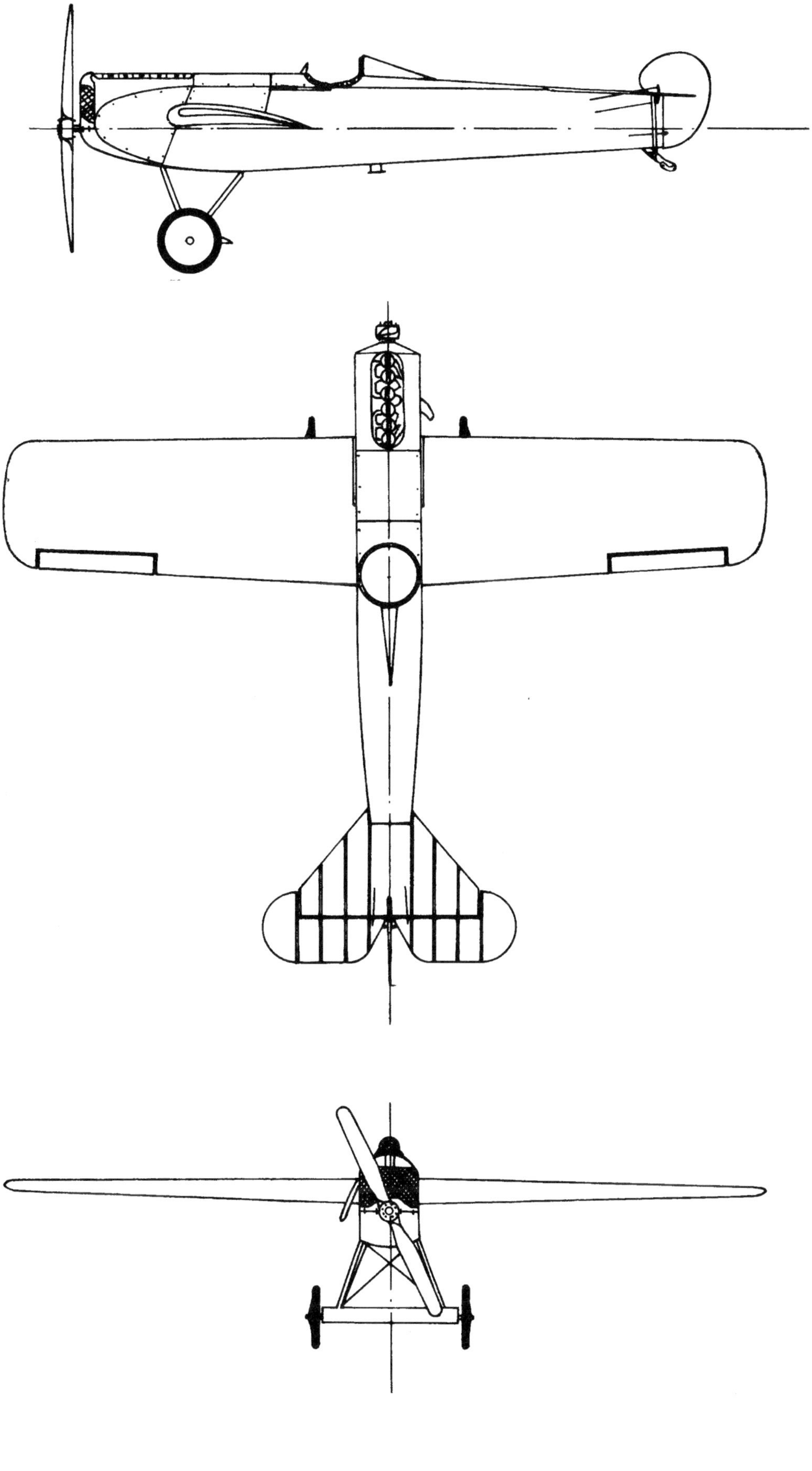

Above & Below: *Idflieg* did not place a production order for the Fokker V23 for the same reason the V20 was rejected, officially because of the poor downward view for the pilot due to the wing location. Power was a 160 hp Mercedes D.III.

Above: Fokker V23 wood cantilever wing under construction. The Fokker V23 was part of a series of cantilever monoplane designs; wings for the V17 and V20 were similar. For reason of impaired field of view, no production was undertaken until the wing was moved to a parasol location with the V28, which was put into production as the E.V and D.VIII after the 'E' designation was cancelled as obsolete. For reasons of aerodynamic drag and strength, the wing was returned to the fuselage for nearly all designs by the 1930s. (Peter M. Grosz Collection/STDB)

Fokker V24

The Fokker V24, Works Number 2612, was a version of the D.VII with a heavier, more powerful, over-compressed 200 hp Benz Bz.IVü engine. After acceptance flights by pilot Gross on 26 April 1918, the V24 took part in the Second Fighter Competition at Adlershof.

This same engine was also tested in the Pfalz D.XIV, a slightly enlarged derivative of the Pfalz D.XII. However, unlike Pfalz, the Fokker V24 had the same wings as the standard Fokker D.VII. The V24 and Pfalz D.XIV were compared with one another during the Second Fighter Competition. In four climb flights between May 27 and June 13, the Pfalz D.XIV, with a starting weight of 1032 kg, averaged 24.2 minutes climb to 5,000m and the V24, at a starting weight of 1006 kg, averaged 22.3 minutes, results that were roughly equivalent.

The additional weight of the Benz Bz.IVü offset the advantage of its greater power, and both performance and maneuverability with the Benz Bz.IVü were not as good as with the 185 hp BMW IIIa. Combined with the need for the Bz.IVü for two-seaters, this eliminated further consideration of the Benz engine for fighter use.

After the climb competition with the Pfalz D.XIV, the V24 was converted to a BMW IIIa engine and after conversion was delivered to ZAK on 17 October 1918.

Fokker V24 Specifications

Engine:	200 hp Benz Bz.IVü	
Wing:	Span	8.90 m
	Area	20.2 m^2
General:	Length	7.14 m
	Height	3.15 m
	Empty Weight	750 kg
	Loaded Weight	989 kg
Climb:	2000m	6.0 min
	3000m	8.5 min
	4000m	12.5 min
	5000m	16.3 min

Above: The Fokker V24 was a variant of the Fokker D.VII powered by the over-compressed 200 hp Benz Bz.IVü engine. Although offering more power, the engine was also significantly heavier, and the 185 hp BMW IIIa engine provided both better maneuverability and better performance due to its lighter weight.

Fokker V 24

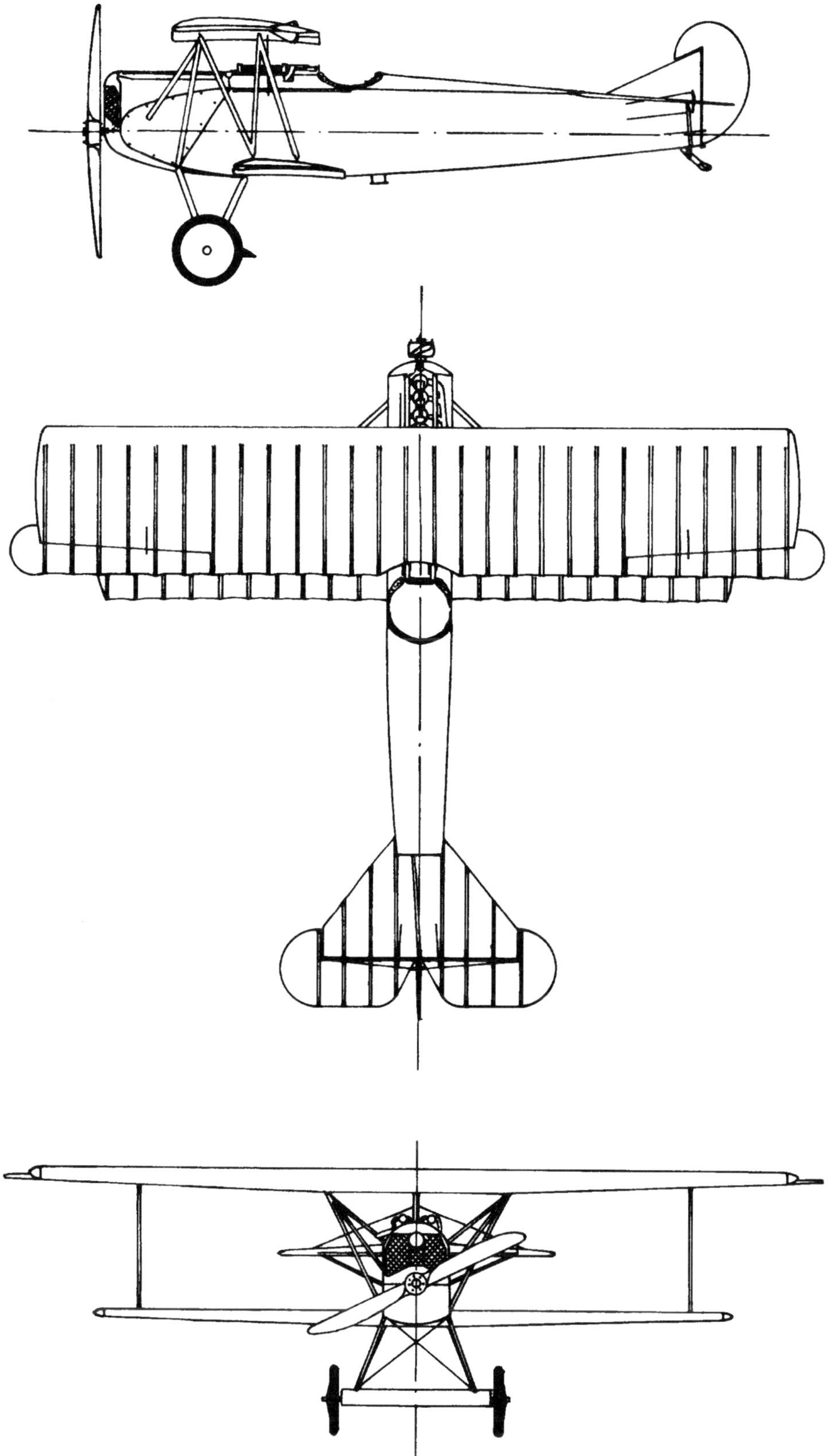

Above & Below: Front and rear quarter views of the Fokker V24 at Schwerin before application of national insignia. Powered by the over-compressed 200 hp Benz Bz.IVü engine, it looks like a standard D.VII with more robust cowling.

Above & Below: Front and rear quarter views of the Fokker V24 at Adlershof after application of national insignia. The additional weight negated potential performance gains from the additional power.

Above: The Fokker V24 powered by the over-compressed 200 hp Benz Bz.IVü engine at Adlershof. The standard D.VII powered by the 185 hp BMW IIIa was lighter and offered better performance and maneuverability.

Above: Closeup view of the radiator of the Fokker V24.

Above: Benz Bz IVaü engine in a test rig.
Right: Benz advertisement.
Below: The Benz Bz.IV engine in the NMUSAF. (NMUSAF)

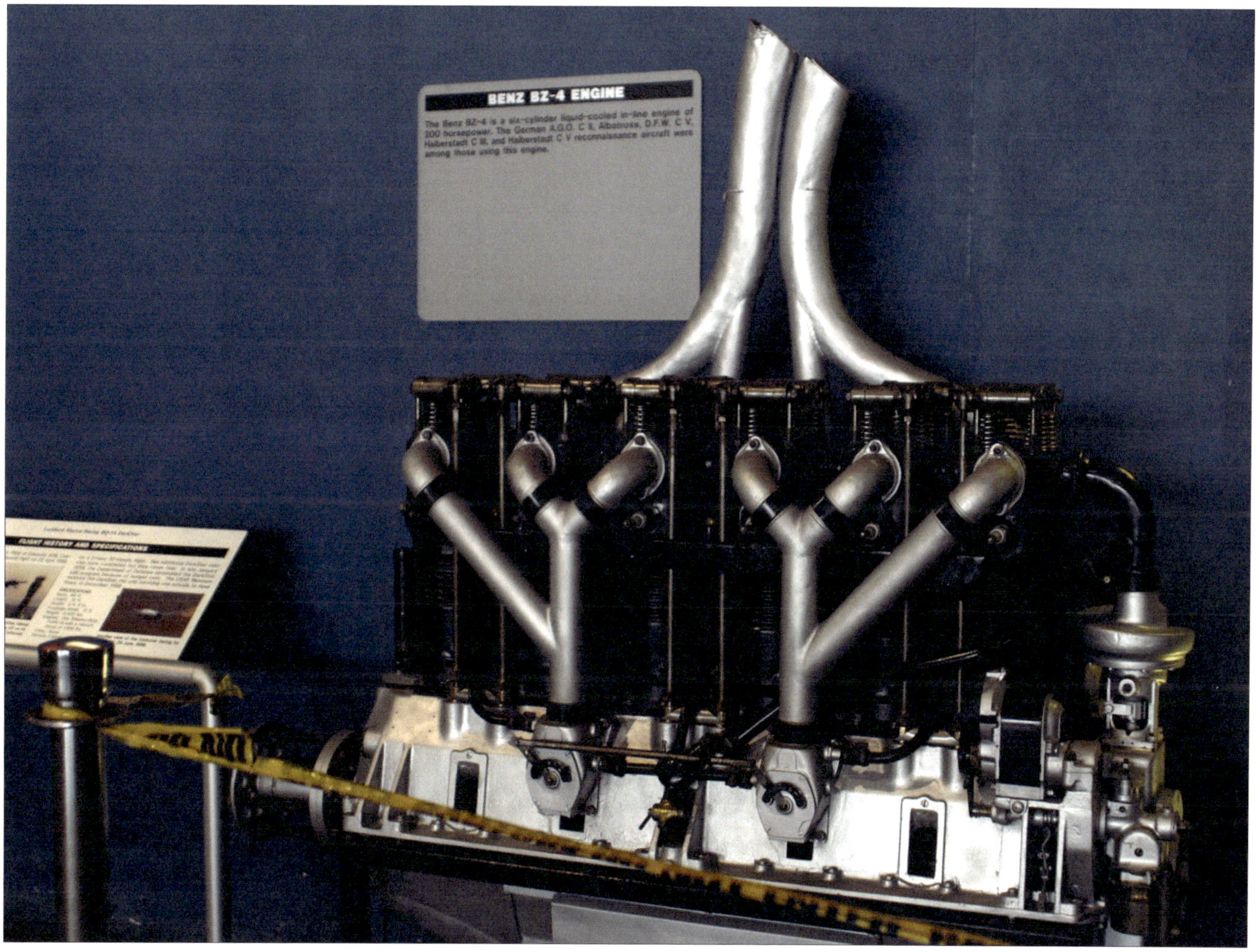

Fokker V25

The Fokker V25, Works Number 2732, was derived from the V17 that had demonstrated such excellent performance at the First Fighter Competition. The V17 wing was mounted in the middle of the fuselage, and to improve the pilot's downward view the V25's wing was mounted at the bottom of the fuselage.

The V25 was entered in the Second Fighter Competiton with the 110 hp Oberursel UR.II and was also tested with the more powerful 170 hp Goebel Goe.III engine. Even the 110 hp engine gave it good performance, but again the pilot's field of view was not considered satisfactory and it was not developed further.

Dimensions of the V25 with Goebel engine were the same as with the Oberursel but the Goebel engine increased the empty weight to 396 kg and loaded weight to 661 kg. Unfortunately, performance with the Goebel engine is not available. After being overhauled at Fokker, the V25 was sent again to ZAK III at Johannisthal on 17 October 1918.

Fokker V25 Specifications

Engine:	110 hp Oberursel UR.II	
Wing:	Span	8.13 m
	Area	9.7 m^2
General:	Length	5.89 m
	Height	3.02 m
	Empty Weight	365 kg
	Loaded Weight	565 kg
Maximum Speed:		200 km/h
Climb:	1000 m	1.5 min.
	2000 m	3.3 min.
	3000 m	6.5 min.
	4000 m	9.5 min.
	5000 m	13.3 min.

Above & Below: The small Fokker V25 with 110 hp Oberursel UR.II engine.

Fokker V 25

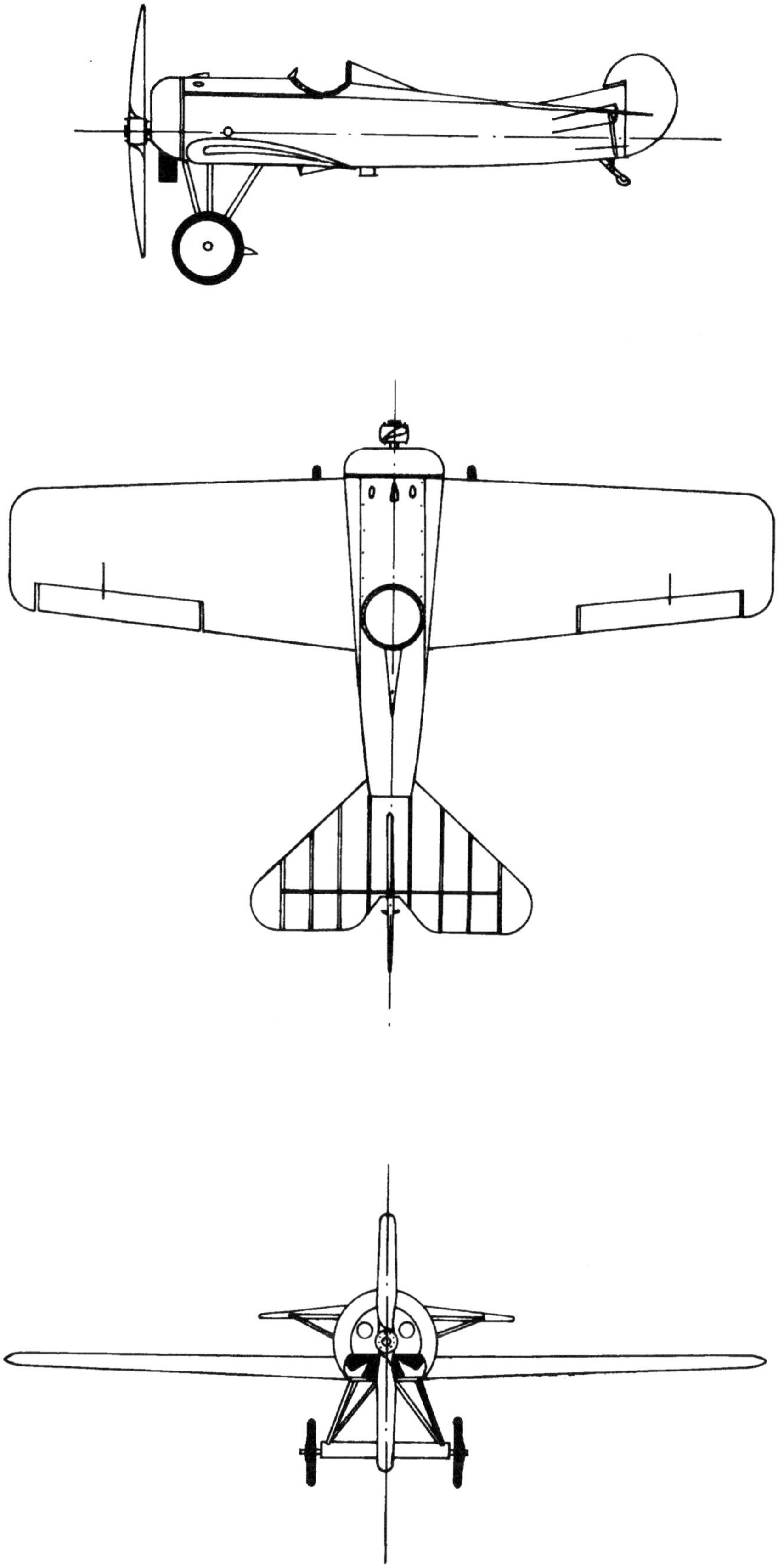

1313.

1311.

Above & Facing Page: The Fokker V25 with 110 hp Oberursel UR.II engine competed at the Second Fighter Competition but, like the earlier V17, despite good performance and maneuverability it was rejected due to the restricted downward field of view for the pilot. A robust crash pylon was built into the pilot's headrest to protect him in case of flipping over on landing. One cannot help but wonder if this rejection echoed an old bias against the E.I–E.IV? Even though the engine of the V17 and V25 offered just 10% more power than that of the E.III, the cantilever wing and cleaner design of the 1918 prototypes offered an overall performance that could match that of any frontline fighter at the time.

Right: The Fokker V25 without wing, cowling, or propeller.

Fokker V26

There is almost no record of the V26 in Fokker or *Idflieg* orders documented at the time, so its identity is as yet unclear. It is possible that the V26 carried Works Number 2733, which lies between the V25 (Works Number 2732) and V27 (Works Number 2734). Perhaps the V26 was the machine that was used in the period from June 7 to 12, 1918, as part of the *Typenprüfung* (Type Test).

Fokker V27

The Fokker V27, Works Number 2734, was another attempt to achieve the performance benefits of the monoplane configuration while solving the pilot's field of view problem that had led to rejection of the earlier Fokker monoplane fighter prototypes. In the V27, the wing was not mounted on the fuselage, which minimized aerodynamic drag, but was moved above the fuselage just above the pilot's eye level, giving the pilot exceptional fields of view both above and below the wing.

The V27 was powered by the experimental 195–200 hp Benz Bz.IIIbo ungeared V-8 engine. It had good performance but the experimental V-8 engine was not yet reliable enough for production and the V27 remained a prototype. Built in April 1918, the V27 was sent to the Second Fighter Competition and was assessed as the best fighter at the competition using the 200 hp Benz Bz.IIIbo. However, vibration prevented production of this engine to the Armistice.

The V27 was shipped to *Idflieg* department A3 after it was repaired.

Interestingly, the Fokker V29, one of the winners of the Third Fighter Competition, used the same configuration as the V27; the main difference was that the V29 used the reliable, production BMW IIIa engine instead on an experimental Benz V-8 engine.

Low drag was the key benefit of cantilever monoplane wings, and the parasol configuration combined that with exceptional field of view. The drawback was that the cantilever wing, even if strong enough to safely support the aerodynamic loads, was not as stiff in torsion as the biplane wing cellule with N-struts, with the result that aileron effectiveness was not as good as that of a biplane.

Fokker V27 Specifications

Engine:	195–200 hp Benz Bz.IIIbo	
Wing:	Span	9.68 m
	Area	14.3 m^2
General:	Length	6.34 m
	Height	3.0 m
	Empty Weight	602 kg
	Loaded Weight	840 kg
Maximum Speed:		200 km/h

Above: The Fokker V27 parasol monoplane configuration solved the problem of the pilot's limited field of view suffered by the previous Fokker monoplane prototypes at the expense of greater drag from the struts supporting its wing. However, despite good performance its 195–200 hp Benz Bz.IIIbo experimental V-8 was not ready for production.

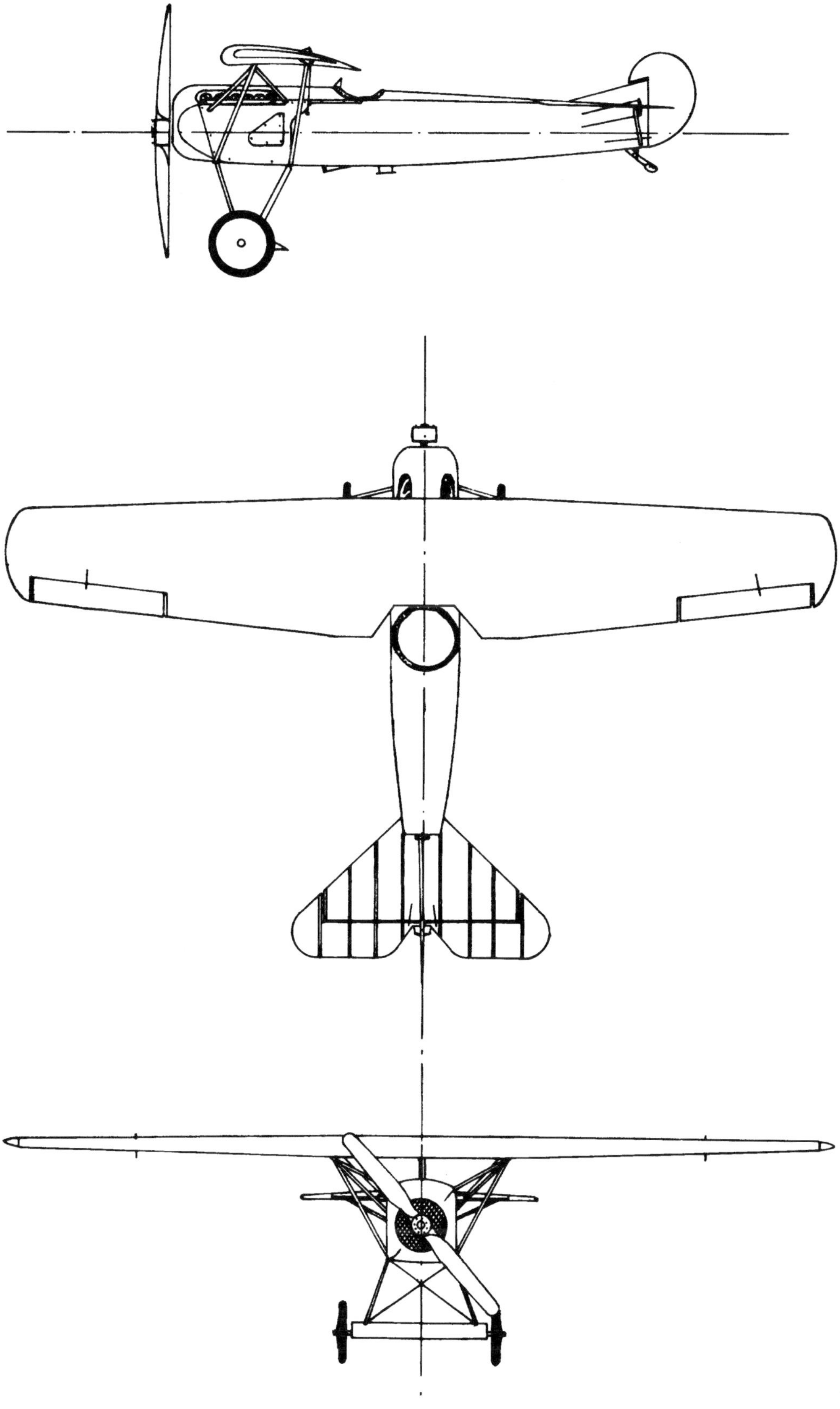

Above & Below: The Fokker V27 parasol monoplane configuration solved the problem of the pilot's limited field of view suffered by the previous Fokker monoplane prototypes at the expense of greater drag from the struts supporting its wing. However, despite good performance its 195–200 hp Benz Bz.IIIbo experimental V-8 was not ready for production.

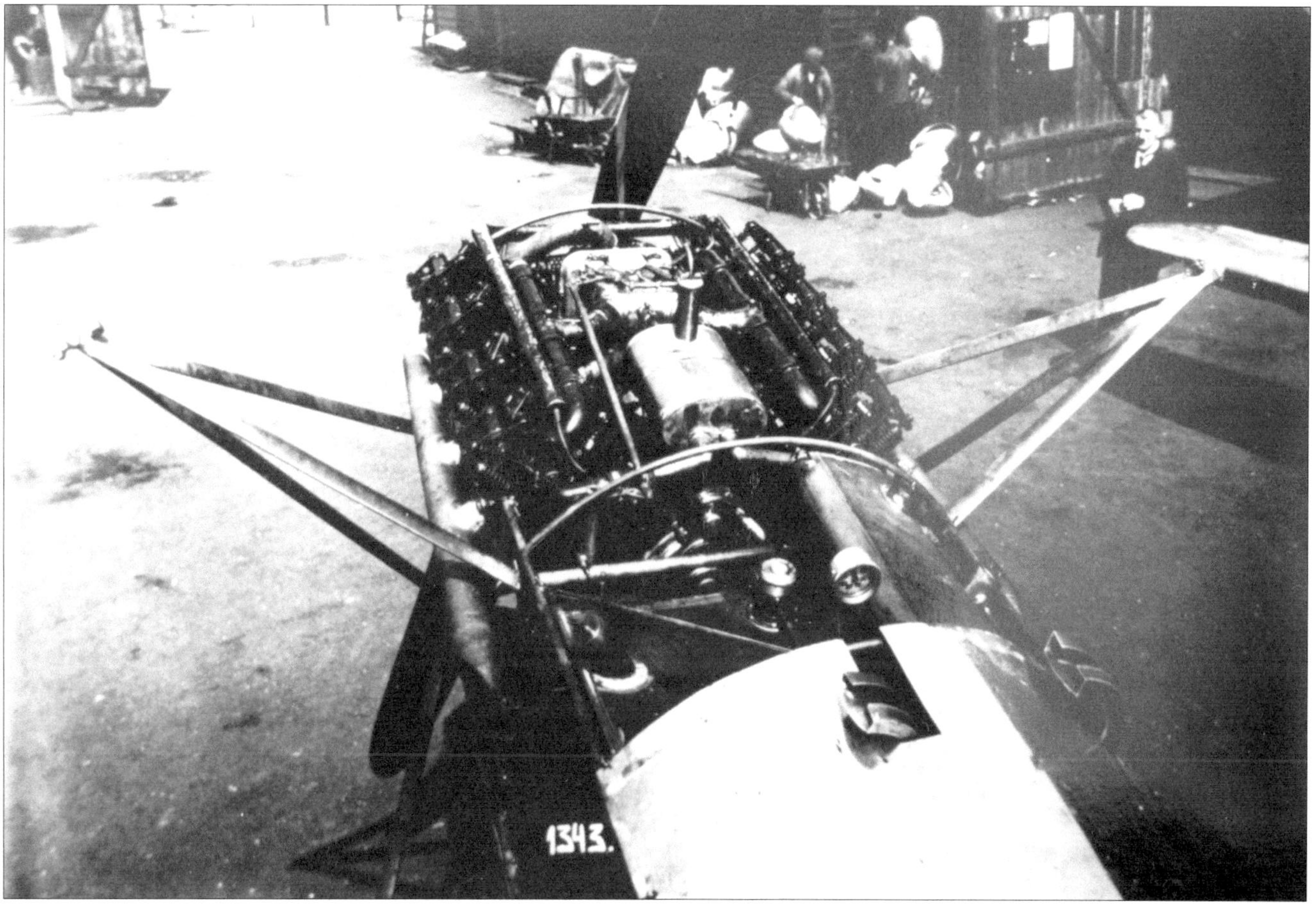

Above: Closeup of the 195–200 hp Benz Bz.IIIbo experimental V-8 engine installation of the Fokker V27 parasol fighter.

Below: A closeup view of the first V27 photo allows a good look at the large radiator required for the V-type engine. The varnish that coated the plywood wing was quite glossy, and the reflection of the unpainted metal forward upper fuselage covering is seen well.

Fokker V28

With the Fokker V28, Works Number 2735, Fokker finally produced another design destined for production as the Fokker E.V. The V28 shared its parasol monoplane configuration with the V27, but was a smaller, lighter machine that was tested with a variety of rotary engines. Its parasol wing solved the pilot's field of view problem that had led to rejection of the earlier Fokker monoplane fighter prototypes at the cost of somewhat increased drag. Like the V27, the wing was moved above the fuselage just above the pilot's eye level, giving the pilot exceptional fields of view both above and below the wing.

The V28 was tested with three different engines during the Second Fighter Competition; the 110 hp Overursel Ur.II, the 145–160 hp Oberursel Ur.III, and the 160 hp Goebel Goe.III. At this time only the Oberursel Ur.II was in production; the other engines were still in development. The V28 was the winner of the Second Fighter Competition despite its low-power 110 hp Oberursel Ur.II, the same engine that had powered the Dr.I Triplane. Despite its low-power the V28 was fast and maneuverable, and had an exceptional field of view for the pilot.

From May 27 to 29, the V28 equipped with the Ur.II engine recorded climb times between 19.3 and 21.6 minutes to 5000 meters. On June 6, the V28 equipped with Ur.III reached 5000 meters in 13.4 to 14 minutes. From June 8 to 12, the V28 equipped with the Geo.III engine reached 5000 meters in 10.4 to 15.7 minutes.

A second V28 [W/N 2738] fuselage went to the *Flugzeugmeisterei* for strength testing on 17 October 1918. A third V28 [W/N 3860] with the Sh.III Rhemag motor took part in the Third Fighter Competition in October at Adlershof.

Fokker V28 Specifications

Engine:	110 hp Oberursel UR.II	
	145–160 hp Oberursel UR.III	
	160 hp Goebel Goe.III	
	160 hp Rhemag Sh.III	
Wing:	Span	8.34 m
	Area	10.7 m^2
General:	Length	5.865 m
	Height	2.82 m
	Empty Weight	360 kg
	Loaded Weight	560 kg
Maximum Speed (UR.II):		200 km/h
Climb (UR.II):	3000 m	7.5 min.
	4000 m	10.5 min.
	5000 m	14.7 min.
	6000 m	19.5 min.

Above: The Fokker V28 parasol monoplane configuration solved the problem of the pilot's limited field of view suffered by the previous Fokker monoplane prototypes at the expense of greater drag from the struts supporting the wing. Smaller and lighter than its companion V-8 powered V27, the V28 was the prototype of the production Fokker E.V that went into production with the 110 hp Oberursel UR.II previously used in the Fokker Triplane and Fokker D.VI biplane. Here the V28 is seen with the 11-cylinder UR.III of 145 hp (some sources give the power as 160 hp, but 145 hp is more likely).

Fokker V 28

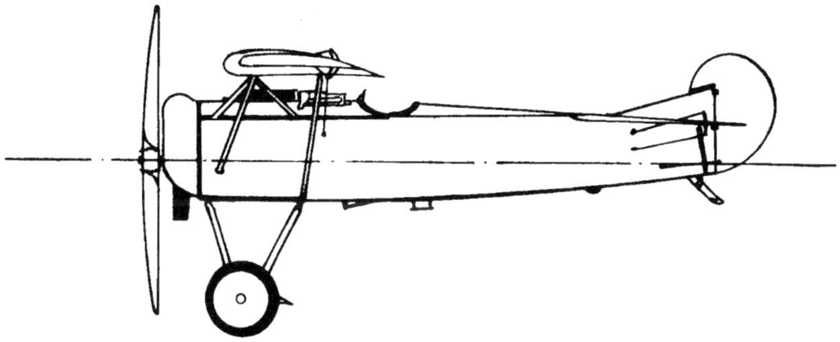

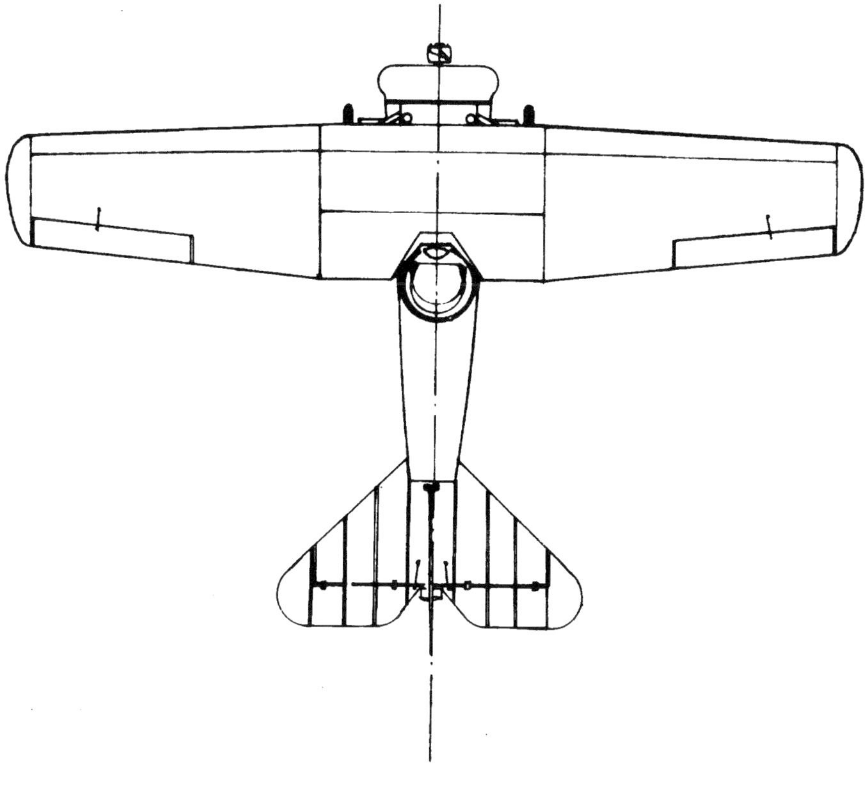

Above: Front view of the Fokker V28 parasol monoplane fighter prototype. The engine appears to be a prototye of the 145 hp Oberursel Ur.III.

Facing Page: The Fokker V28 parasol monoplane fighter prototype, test pilot, and observers. The engine appears to be a prototye of the 145 hp Oberursel Ur.III.

The first mention of an unbraced Fokker monoplane with a 160 hp Göbel engine is in the April report of *Idflieg*, which notes that the aircraft climbed to 5000 m in 11 minutes according to factory data. Since the use of the Goe.III rotary engine in a monoplane is so far only known for the V28, it must have been this model, although no further information was found on the attempt.

Already in the next *Idflieg* monthly report in May, it was decided to build the Fokker Parasol monoplane with 110 hp Ur.II engine as the E.V instead of the 60 D.VIs not yet delivered from the order for 120.

1338.

Above & Below: The Fokker V28 parasol monoplane fighter. Here the V28 is seen with the 9-cylinder Ur.II of 110 hp. The wing tips are more squared off than the E.V/D.VIII production aircraft.

Above & Below: The Fokker V28 parasol monoplane fighter prototype.

Above: The Fokker V28 parasol monoplane fighter. Here the V28 is seen with the 11-cylinder Ur.III of 145 hp.

Below: The wingless V28, armed with only the left machine gun, was fitted with a "bullet deflector" prop for trials. This was a copy of the propeller fitted to early Morane-Saulnier L and n monoplanes – about three years earlier. Why such a system was evaluated by Fokker in 1918 remains speculative. (Peter M. Grosz Collection/STDB)

OBERURSEL
FLUGMOTOREN
MOTORLOKOMOTIVEN
MOTORENFABRIK OBERURSEL A.G.
OBERURSEL

Fokker=Werke
G.m.b.H.
Schwerin i.M.
Das ideale
Sport=Flugzeug.=

Oberurseler Umlaufmotor
für Eindecker
Doppeldecker
Kampfflugzeuge
Motorenfabrik Oberursel A.G.
Oberursel

FOKKER
SCHWERIN, MECKL. BERLIN-REINICKENDORF
LUFTWAFFEN

Fokker V29

Above: The Fokker V29 shown here was one of the two winners of the Third Fighter Competition. Essentially, it was a parasol monoplane derivative of the Fokker D.VII. Although similar in configuration to the smaller, rotary-powered D.VIII, it had the more powerful 185 hp BMW IIIa engine for better high altitude performance. Similarly, the more powerful D.VII biplane had eclipsed the smaller, rotary-powered D.VI of similar configuration.

After winning the first two fighter competitions, Fokker entered the Third Fighter Competition in the best position of any manufacturer. The Fokker D.VII was the mainstay of the German fighter units and the BMW-powered version was generally acknowledged as the best all-around fighter in combat. Fokker's D.VII was being built not only by Fokker but by two other manufacturers under license. After a rocky start due to two fatal wing failures, Fokker D.VIII production was back in full swing.

Fokker's entries in the third competition built on these successes. The V28 was a D.VIII prototype with the more powerful 160 hp Seimens-Halske rotary, the V36 was a modified D.VII, and the V29 (Works Number 2736) was a parasol monoplane derivative of the D.VII that was a high-altitude fighter.

As early as June 1918 the V29 was mentioned in the *Idflieg* monthly report as it was undergoing flight testing equipped with the BMW IIIa engine. In July *Idflieg* stated that, according to the results of the flight tests, the aircraft in its present form was not yet suitable as a replacement for the Fokker D.VII with BMW engine. The V29 was tested with two different wings of 13.3 and 14.3 square meters wing area.

Fokker V29 Specifications

Engine:	185 hp BMW IIIa	
Wing:	Span	9.68 m
	Area	14.3 m^2
General:	Length	7.02 m
	Height	2.93 m
	Empty Weight	632 kg
	Loaded Weight	861 kg
Maximum Speed:		200 km/h
Climb:	3000m	7.5 min
	4000m	10.5 min
	5000m	14.7 min
	6000m	19.7 min

Fokker V 29

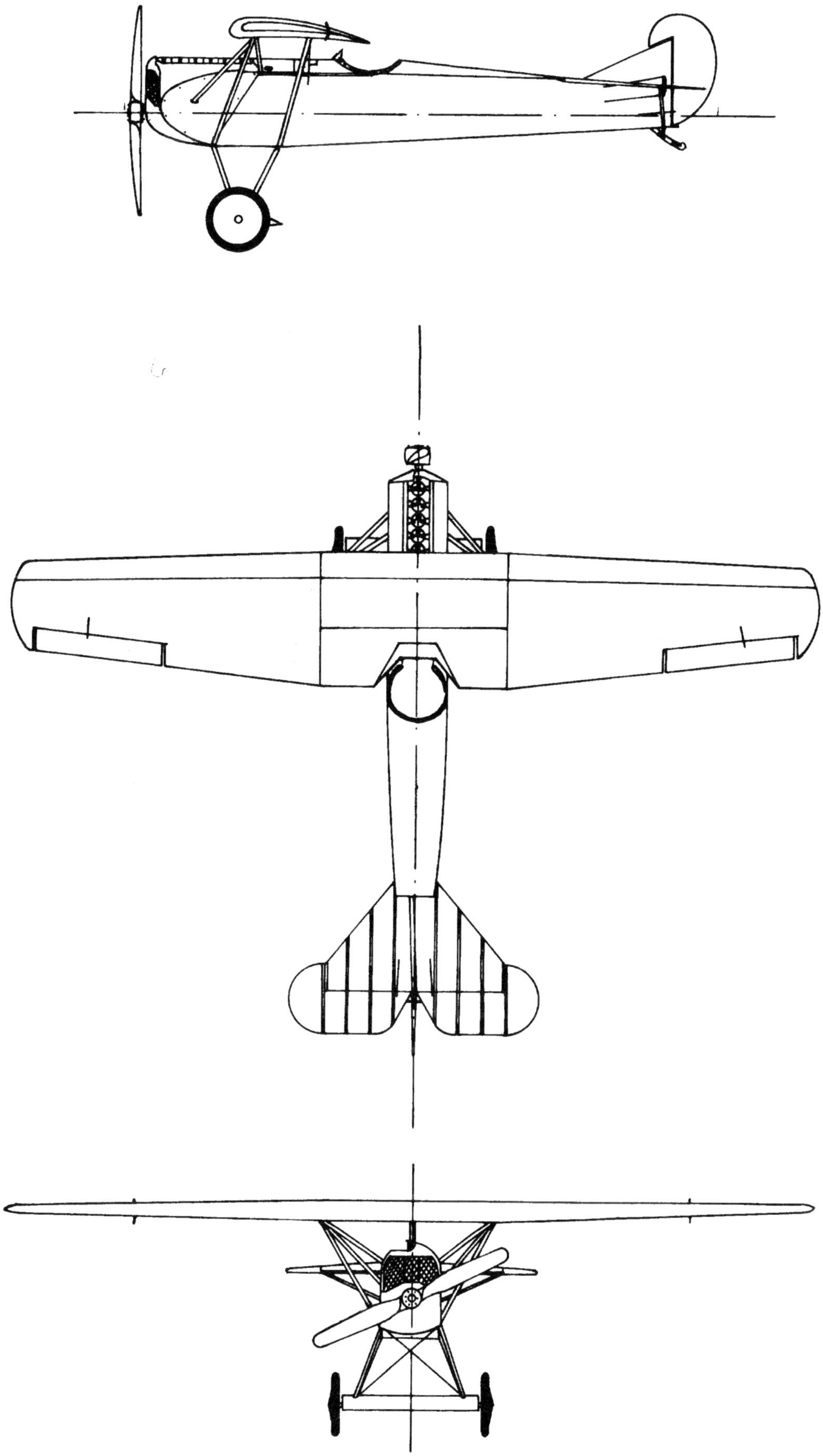

Using the same basic configuration as the production D.VIII, the V29 was evaluated at the Third Fighter Competition with both the over-compressed BMW IIIa and Mercedes D.IIIaü engine for excellent high-altitude performance and was one of the winners of the competition, the other being the Rumpler D.I. However, with fresh memories of the D.VIII wing debacle and with the robust, proven D.VII still superior, *Idflieg* was in no hurry to order the V29 into production and had not done so by the Armistice. Similarly, the troubled Rumpler D.I never reached mass production.

Above & Below: The Fokker V29 was one of the two winners of the Third Fighter Competition. It was evaluated with both the more Mercedes D.IIIaü and the 185 hp BMW IIIa engine at the competition for better high altitude performance. The Rumpler D.I was the other winner of the competion, but Rumpler could never bring this troubled design to a configuration ready for mass production.

Fokker V30

The V30 was a cantilever parasol monoplane (w/n 2737) with a plywood wing powered by Benz Bz.IIIbm geared V-8 motor of 195–200 hp.

The V30 was sent to Adlershof on July 25, 1918 for evaluation and went there again in October for strength testing. According to the Fokker factory, the V30 was also tested with two different sized wings of 13.3. and 14.3 square meters wing area.

No photo of the aircraft has been found. Its description sounds much like the V27 only powered by the geared version of the experimental Benz V-8

Fokker V30 Specifications

Engine:	195–200 hp Benz Bz.IIIbm V-8	
Wing:	Span	9.68 m
	Area	13.3 m^2
General:	Length	6.535 m
	Height	3.075 m
	Empty Weight	652 kg
	Loaded Weight	882 kg
Maximum Speed:		about 200 km/h
Duration:		1.5 hr

Fokker V31

A Fokker document "New Constructions and Test Machines" written during the war refers to the V31 as a two-seat parasol monoplane with wing area of 16.5 square meters and 185 hp BMW IIIa engine. No more data has been found.

The V31 sounds like it may have been an enlarged, two-seat derivative of the V29.

Right: Seen in the background of a Fokker V34 photo is this parasol monoplane. It may be the Fokker V31.

Fokker V32

No documented information is known.

Fokker V29 prototype, co-winner of the Third Fighter Competition.

Fokker V33

Above: The Fokker V33 was derived from the V9 prototype. It was tested with both the 110 hp Oberursel Ur.II and the 145 hp Oberursel Ur.III. Its high altitude climb rate did not match the BMW-powered Fokker V.29.

The Fokker V33 was a small, rotary-powered biplane fighter derived from the Fokker V9 that was tested at Schwerin in August 1918. It was tested with both the 110 hp Oberursel Ur.II (a 9-cylinder engine) and the 145 hp Ur.III (an 11-cylinder engine). Both versions had the same dimensions, but the Ur.III-powered version was heavier due to its heavier engine. The empty weight of the Ur.III-powered version was 396 kg and the loaded weight was 616 kg.

Fitted with the 110 hp Ur.II, the V33 climb rate approximated that of the more powerful V29 up to 4000 meters, but fell behind above that altitude. Unfortunately, data for the V33 climb rate when fitted with the 145 hp Ur.III are not available.

After the war, Fokker flew the V33 in Holland. Dutch newspaper reports mention aerobatic demonstrations of what was then described as a sports plane with an 80 hp Le Rhône engine in the summer of 1919.

Fokker V33 Specifications

Engine:	110 hp Oberursel Ur.II	
	145 hp Oberursel Ur.III	
Wing:	Span	7.24 m
	Area	13.7 m^2
General:	Length	5.456 m
	Height	2.385 m
	Empty Weight	360 kg
	Loaded Weight	560 kg
Maximum Speed (Ur.II & Ur.III):		200 km/h
Climb (Ur.II):	3000m	7.5 min
	4000m	11 min
	5000m	17 min
	6000m	24 min
Endurance:		1.5 hrs

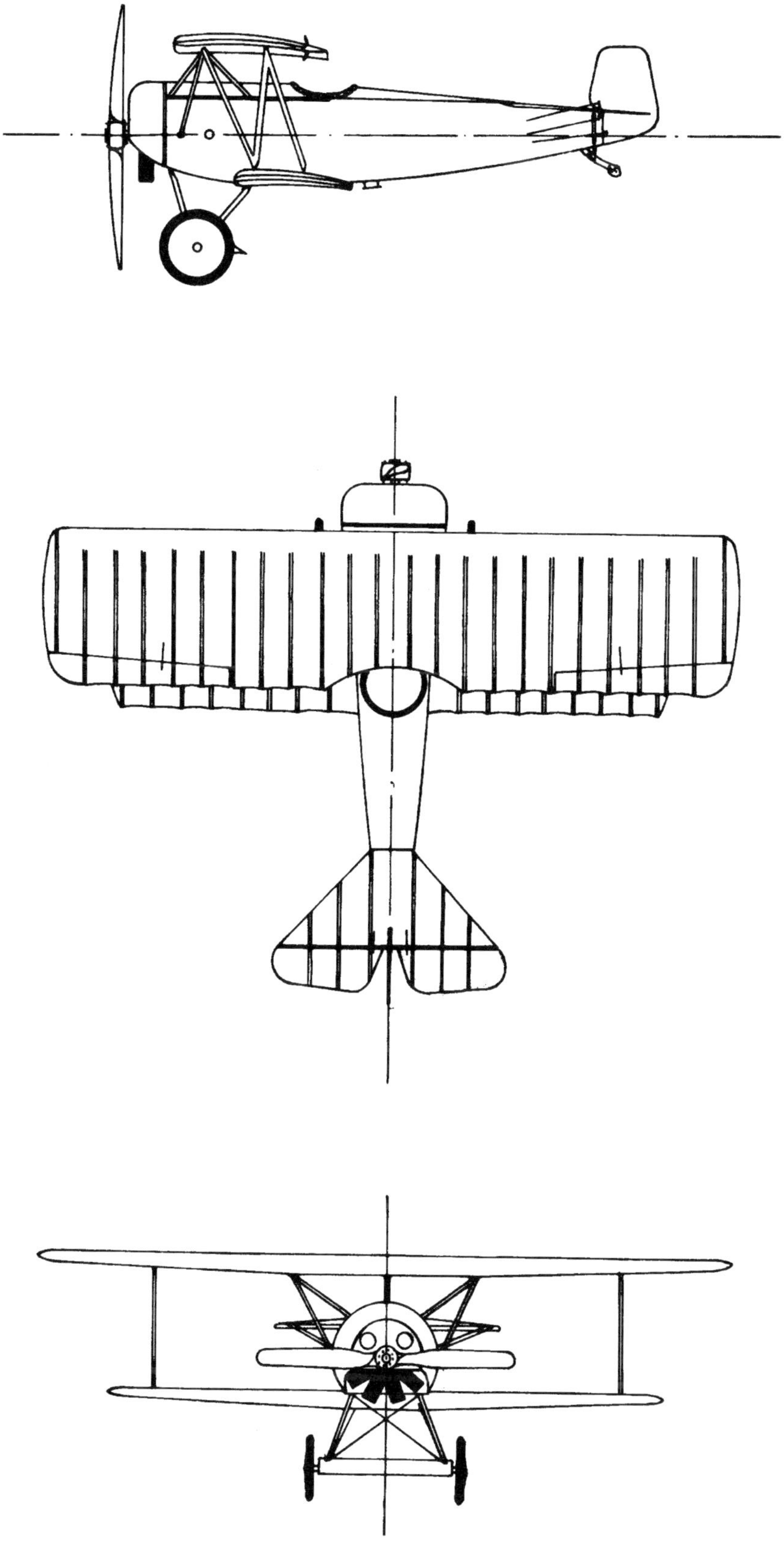

Above: The Fokker V33 was derived from the V9 prototype. It was tested with both the 110 hp Oberursel Ur.II and the 145 hp Oberursel Ur.III. Its rudder was a different shape than the customary Fokker comma shape.

Above: The Fokker V33 was derived from the V9 prototype. It was tested with both the 110 hp Oberursel Ur.II and the 145 hp Oberursel Ur.III. The engine fitted here appears to be the Ur.III.

Right & Below: The Fokker V33 in the Netherlands postwar.

Above & Above Right: The Fokker V33 prototype under construction with the 145 hp Oberursel Ur.III.

Fokker V34

Above: The Fokker V34 in its initial configuration without national insignia. Note the vertical tail surface design. The aircraft in the right background may be the Fokker V31.

The Fokker V34, Works Number 3254, was a refined, slightly smaller development of the Fokker D.VII to improve its performance. Its oval radiator and two distinct designs of rudder are its most easily noticed differences, but perhaps the most important was reduced wingspan and reduced chord upper wing that eliminated the need for a center-section cutout. It was built in September 1918 and on October 10 was sent to participate in the Third Fighter Competition, but it was not selected as the winner, indicating it was not a significant improvement over the standard D.VII, which remained in production until the end of the war. Power was provided by the BMW IIIa, the best inline engine available at this time.

Fokker V34 Specifications

Engine:	185 hp BMW IIIa	
Wing:	Span	8.45 m
	Area	17.6 m^2
General:	Length	6.76 m
	Height	3.01 m
	Empty Weight	628 kg
	Loaded Weight	864 kg
Maximum Speed:		200 km/h

The Fokker V34 in its second configuration with revised vertical tail design and national insignia applied. The fuselage of V34 was covered in four-color fabric.

Fokker V 34

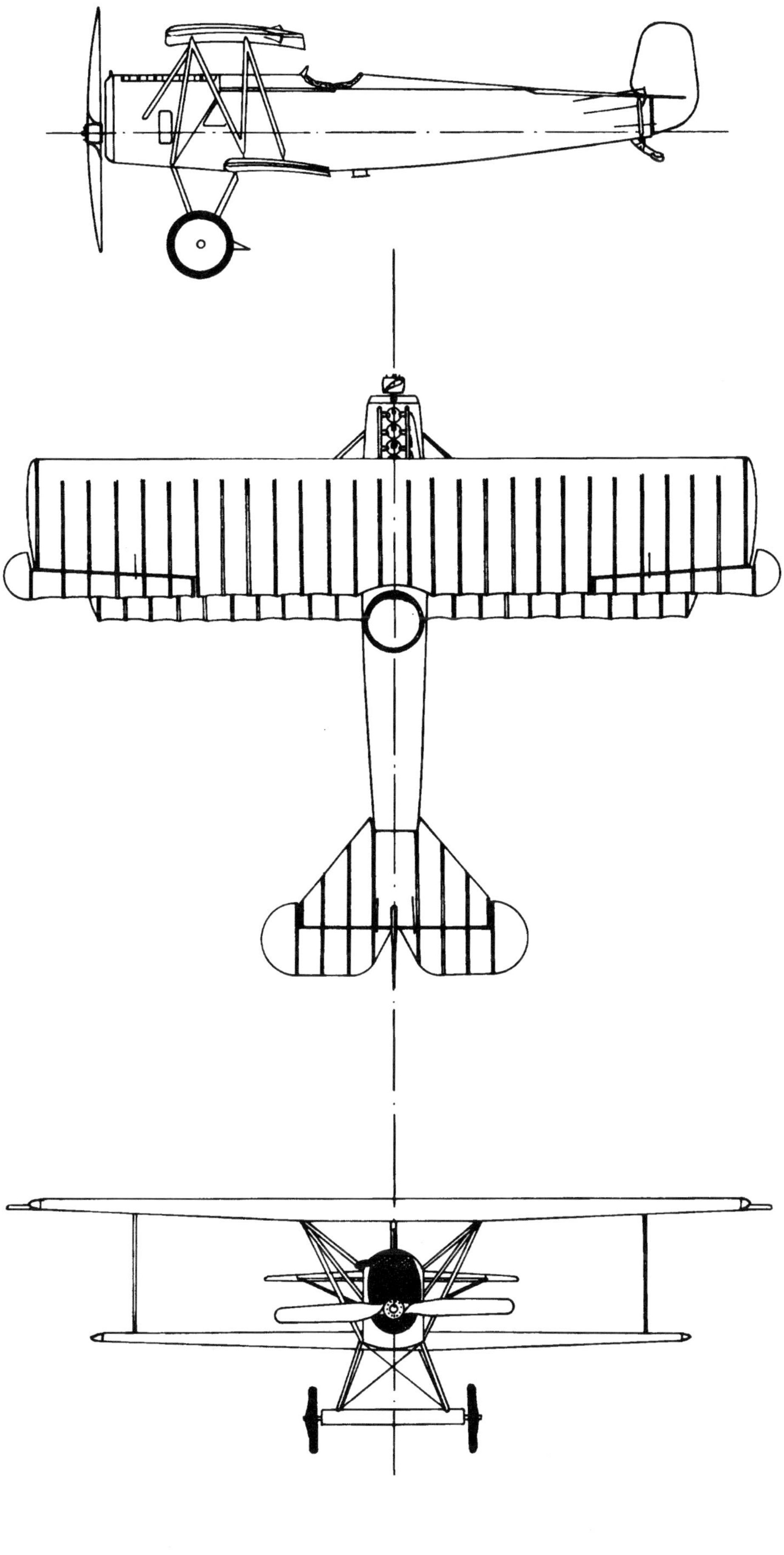

Above & Below: The Fokker V34 was derived from the Fokker D.VII; it had slightly smaller dimensions, an oval radiator, and was powered by the 185 hp BMW IIIa. Its most significant change from the production D.VII was its reduced wing span and reduced chord of the upper wing, which eliminated the need for a center-section cutout for the pilot. Initially it had a sizable fixed fin; its second tail configuration shown here replaced the initial tail with a trapezoidal rudder without fin. The V34 fuselage was covered with four-color camouflage fabric.

Right: The steady increase in production orders for the BMW IIIa engine meant that the company could afford to place full-page color adverts in magazines such as *"Motor"* in 1918.
These present the company logo in its original version, which has since been modernized in appearance several times. The basic logo remained the same for more than one hundred years now, and the white and blue center symbolizes a turning propeller against the Bavarian blue sky.
The success of the first product of the company was so overwhelming that the Third Fighter Competition, held in Berlin in October 1918, was limited to aircraft prototypes powered by the BMW IIIa engine. Aircraft powered by other engines were only included in order to allow for a general comparison of these to the new "standard" inline engine. For a company whose first mass-product was only introduced to frontline service in numbers in June 1918, this was quite an achievement. This was a direct consequence of the successful pairing of the powerful engine to the innovative Fokker D.VII airframe design. And, of course, Fokker wanted to continue this by entering the V34 and other new designs powered by the BMW engine into the trials at Berlin.

Left & Above: The propellers manufactured by the Axial company were fitted to many of the Fokker fighters that were built in the last two years of the war. They show up in the overwhelming number of photos taken of the Fokker Triplane and E.V, as well as on many of the Fokker-built D.VIIs. The company logo was applied as a transfer to their products for a long time, but around the middle of 1918 this practice was discontinued. The prop mounted to V34 was also made by Axial, but also lacked the company decal.

Fokker V35

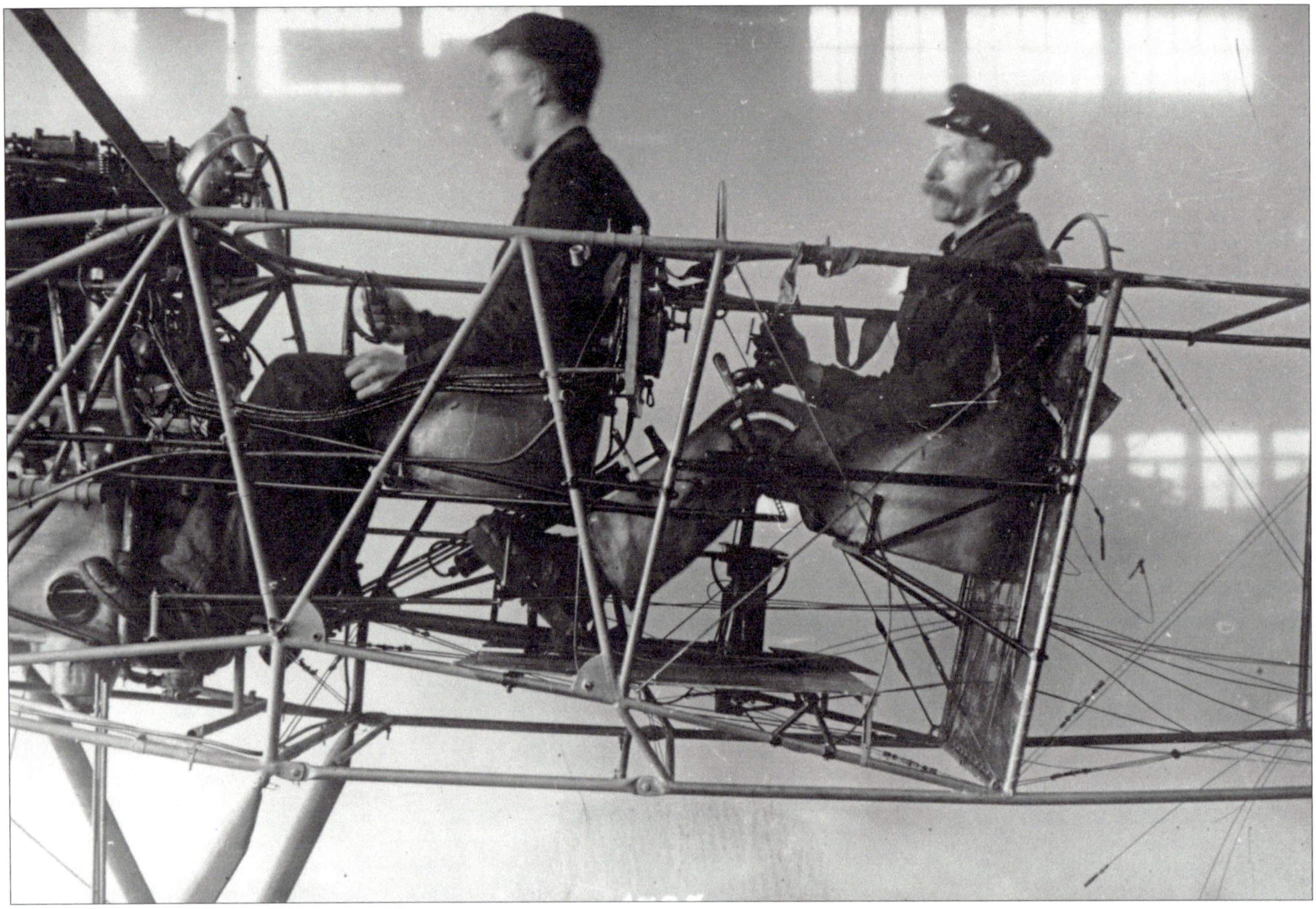

Above: Uncovered fuselage of a two-seat D.VII with crewmembers in place. Dual controls are fitted, making this version suitable as a trainer. This may have been the uncovered V35. (Peter M. Grosz Collection/STDB)

In 1918 the Fokker-Flugzeugwerke produced several variations of the successful production Fokker D.VII fighter. One of these was the V.35 reconnaissance fighter. The machine was a two-seater and was unarmed during the tests (it was planned to mount machine guns later).

The second cockpit displaced the fuselage fuel tank; the fuel was in an airfoil-shaped tank between the wheels. The engine was a 185 hp BMW IIIa.

The V.35 was flown by Ernst Udet postwar, but it did not show any serious advantages over two-seat production fighters.

Above: Uncovered fuselage of a two-seater variant of the Fokker D.VII – that may be V35 – showing the structure and limited space for the rear crew member. The fuel tank was in the thick airfoil between the wheels, and the vertical tubes are the fuel lines. The V35 had an airfoil fuel tank, consistent with this photo. The later V38 had more room for the observer and a flexible gun. (Peter M. Grosz Collection/STDB)

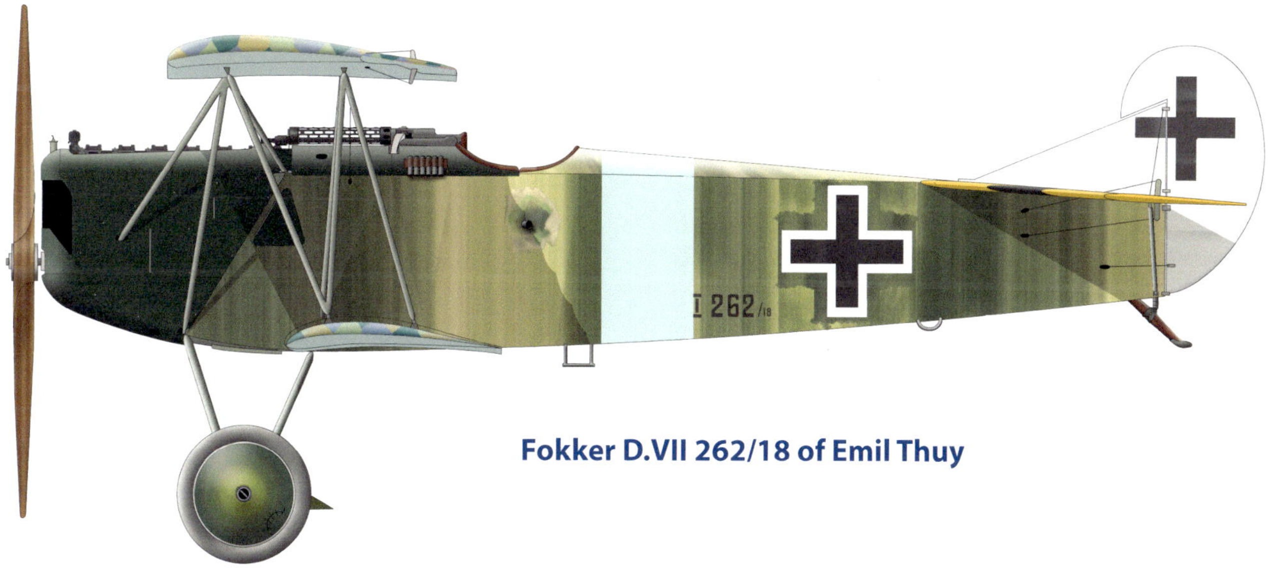

Fokker D.VII 262/18 of Emil Thuy

Fokker V36

Above: The Fokker V36, Works Number 3256 was a development of the BMW-powered Fokker D.VII. The main changes were the oval radiator and fuel tank located in the undercarriage fairing.

Analysis of the information available on the V36, Works Number 3256, indicates that it was as closely related to the V34 as the V18 was to the V11, and the V 27 was to the V 28 and V29. Following the adoption of the cantilever wing, Fokker had obviously become inclined to develop pairs or even trios of prototypes that were closely related to one another.

Visually, the V36 was closely related to the D.VII. The rear fuselage, including the horizontal tail surfaces, fin and rudder were almost identical, but the fuselage was about half a meter shorter compared to the D.VII. The forward fuselage, with its oval engine cowling and radiator, closely resembled that of the V34. The lower wings appeared to retain the layout seen on the D.VII, while the upper wing had a reduced chord, resulting in a smaller wing area.

The fuel tank was relocated into the undercarriage axle wing fairing, which was intended to reduce the fire hazard to the pilot. Like the V34, it was built in September 1918, and together they were shipped on 10 October 1918 to participate in the Third Fighter Competition. Again, detailed information concerning the assessment of the type by the participating pilots is not available, but it was not selected as one of the winning designs.

Fokker V36 Specifications

Engine:	185 hp BMW IIIa	
Wing:	Span	8.935 m
	Area	17.6 m^2
General:	Length	6.46 m
	Height	3.045 m
	Empty Weight	637 kg
	Loaded Weight	871 kg
Maximum Speed:		200 km/h
Climb:	1000m	1.75 min
	2000m	4 min
	3000m	6.75 min
	4000m	10 min
	5000m	13.5 min
	6000m	18.25 min
Endurance		1.5 hr.

Fokker V 36

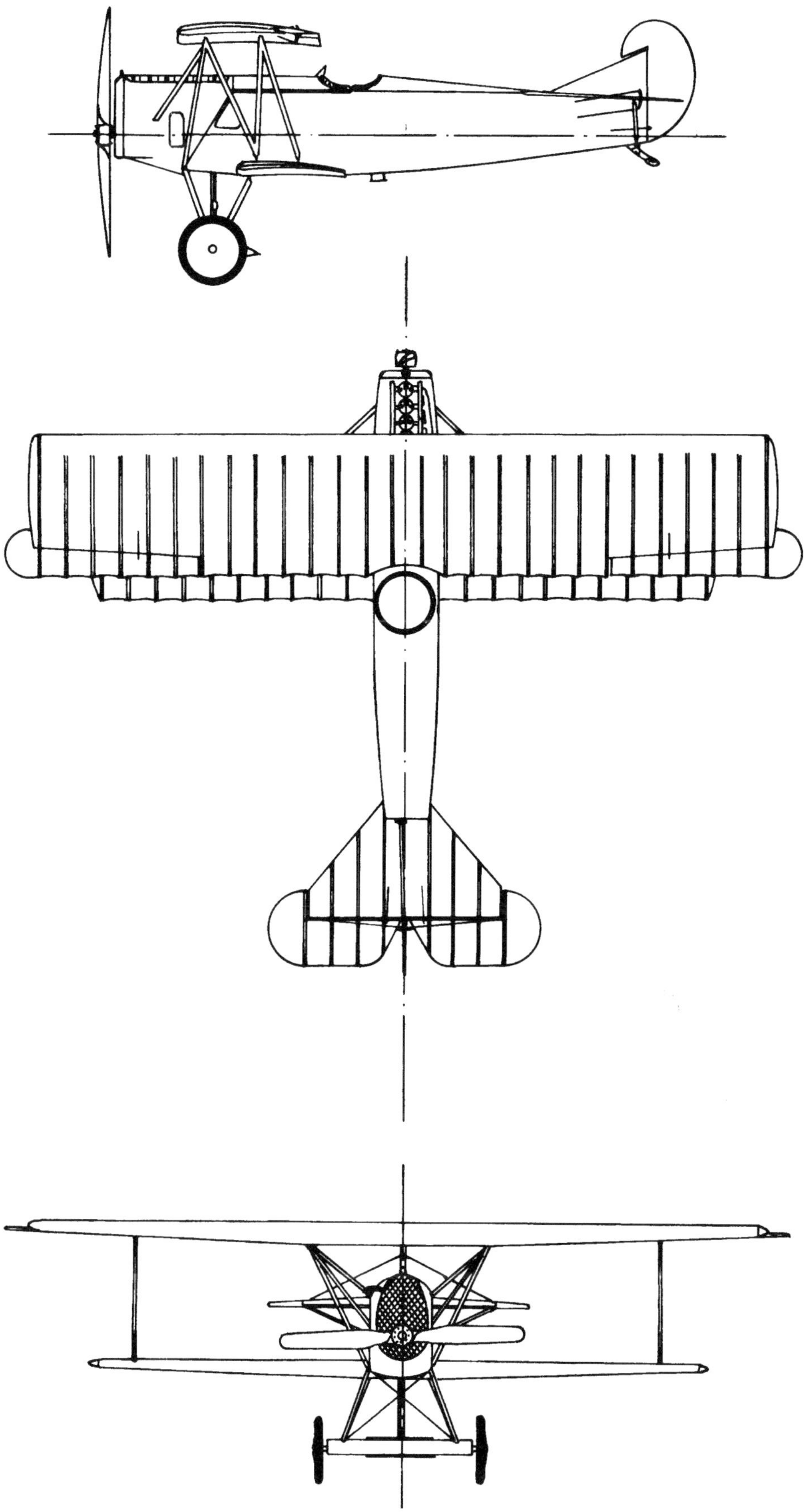

1495.

1498.

Above, Below, & Facing Page: The Fokker V36, Works Number 3256 was a development of the BMW-powered Fokker D.VII and used an oval radiator similar to the V34. The main changes were reduced dimenions, including wing span and chord of the upper wing, the oval radiator, and fuel tank located in the undercarriage fairing. It participated in the Third Fighter Competition that was won by the Fokker V29, a parasol monoplane derivative of the BMW-powered D.VII.

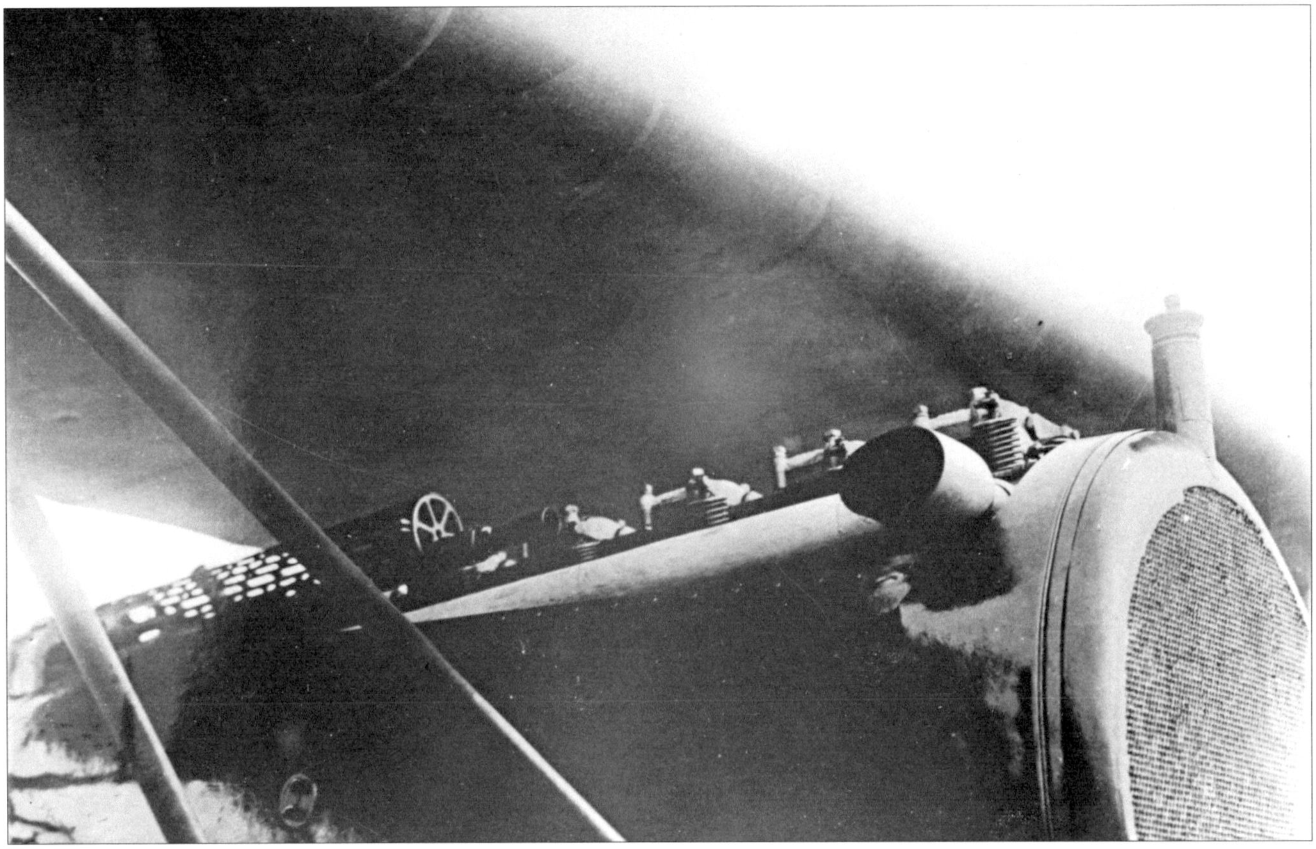

Above: The Fokker V36 engine and radiator closeup. (Peter M. Grosz Collection/STDB)

Above: The Fokker V36 was very similar to the V34; both were derived from the Fokker D.VII, both had an oval radiator, smaller wings, and were powered by the 185 hp BMW IIIa . The V36 differed from the Fokker V34 in its vertical tail and movement of the fuel tank to the undercarriage airfoil. Both V34 and V36 were entered in the Third Fighter Competition that was won by the Fokker V.29.

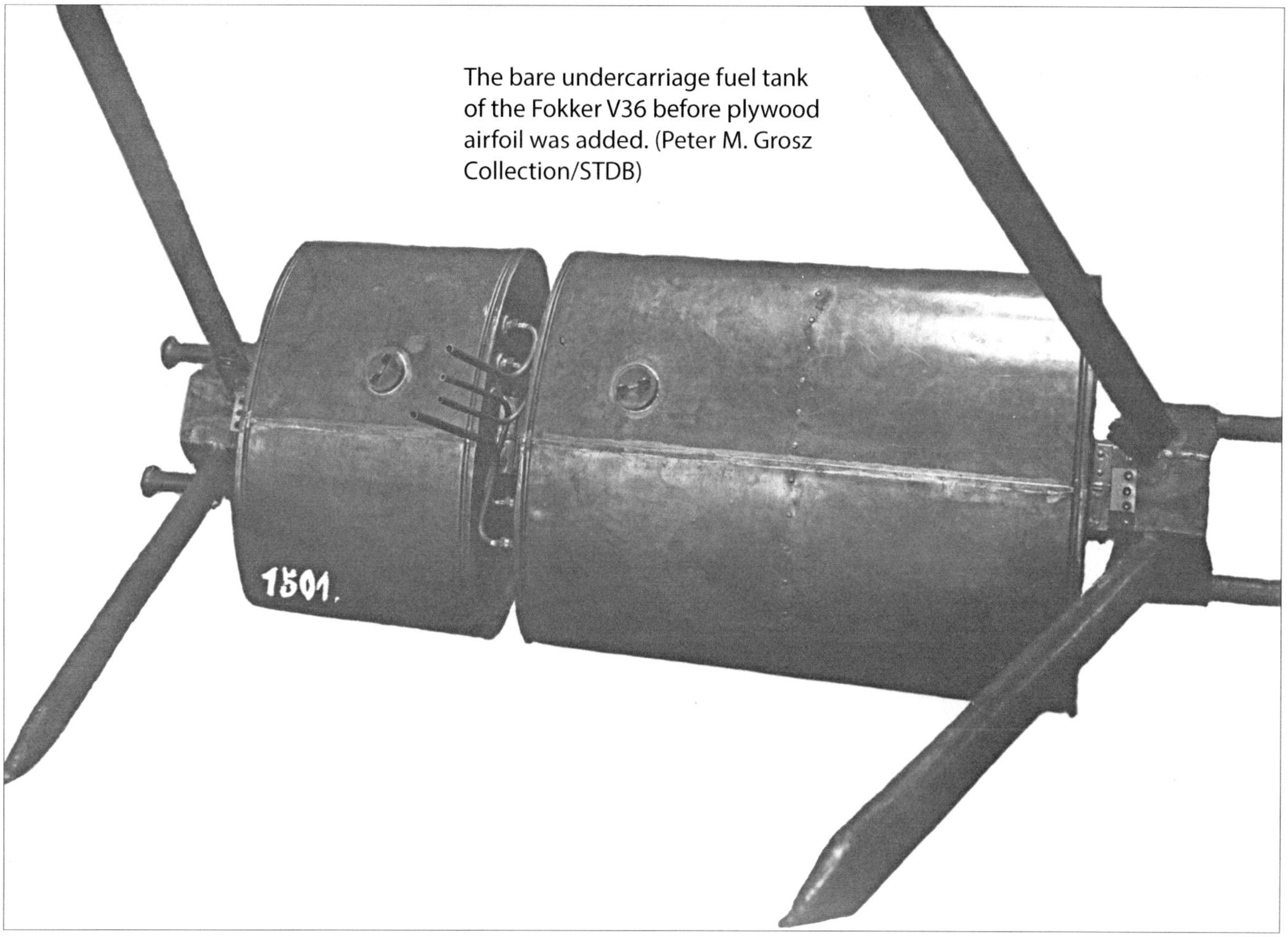

The bare undercarriage fuel tank of the Fokker V36 before plywood airfoil was added. (Peter M. Grosz Collection/STDB)

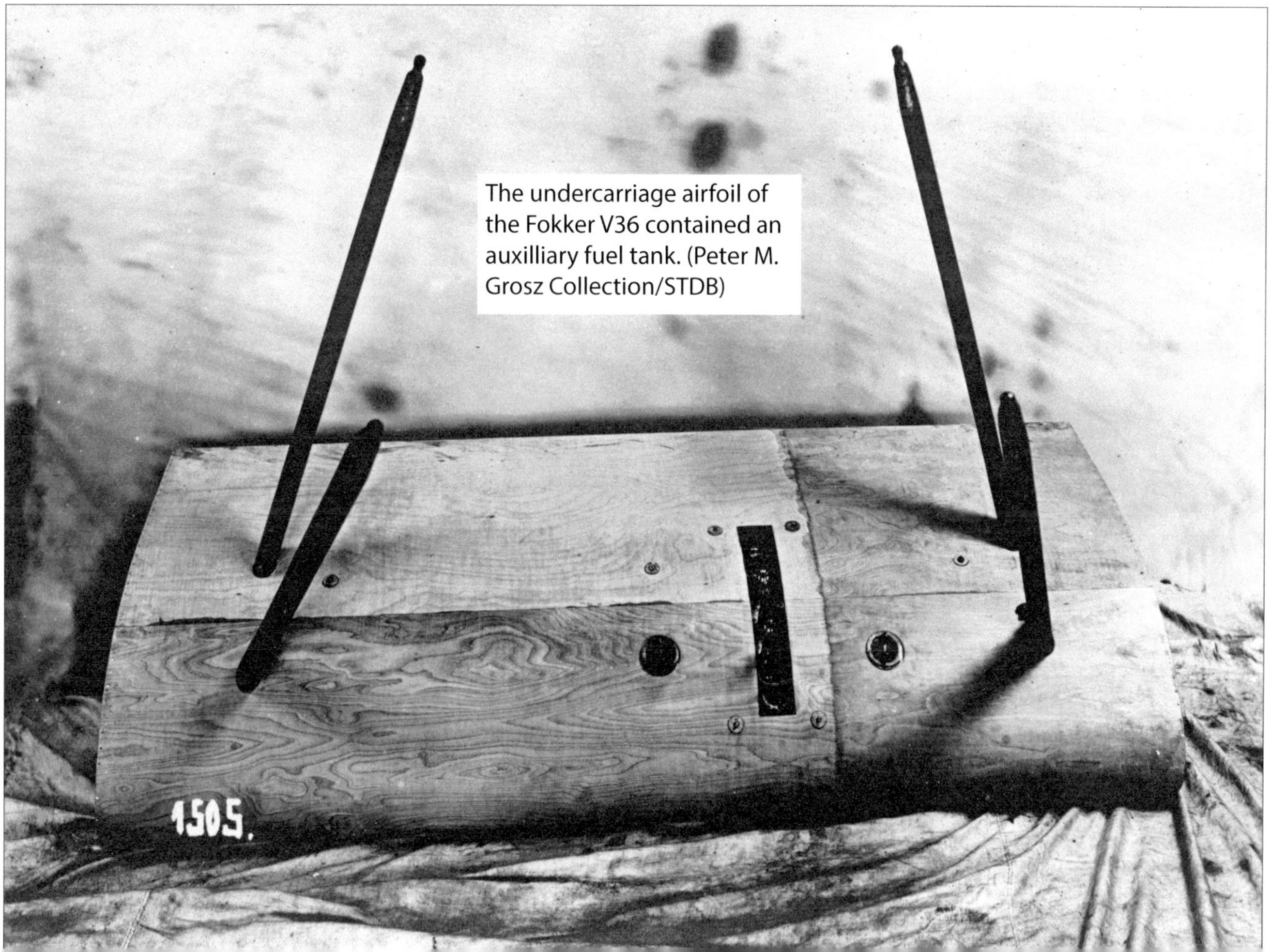

The undercarriage airfoil of the Fokker V36 contained an auxilliary fuel tank. (Peter M. Grosz Collection/STDB)

Fokker V37

Above: The Fokker V37 armored fighter prototype combined the configuration of the earlier Fokker V27 with a distinctive propeller spinner and the 195 hp Benz Bz.IIIb V-8 water-cooled engine. The entire forward fuselage of the Fokker V37 was armored with 2.5mm steel to protect the pilot, engine, and radiator from rifle-caliber ground fire. (Peter M. Grosz Collection/STDB)

Derived from the earlier Fokker V27, and like it powered by a 195 hp Benz Bz.IIIb liquid-cooled V-8 engine, the Fokker V37 was an armored fighter designed for low-level fighting and ground attack.

Clearly the radiator had to be protected from ground fire to make an effective armored airplane, and the V37's solution was clever. The radiator was mounted behind the propeller and protected by a disk of armor. The propeller was given a large spinner for streamlining, and air was forced through the radiator by six fan blades mounted in the plane of the propeller. The spinner was made of aluminum sheet and an armor plate disk. Fokker received Reichs patent No. 335745 on 2 October 1918 for an armored propeller spinner with ventilator-like slots.

The entire forward fuselage, including the engine and cockpit, was protected by steel armor plate 2.5mm thick. The armor behind the pilot even extended up behind his head and was faired into a headrest.

Fokker V37 Specifications

Engine:	195 hp Benz Bz.IIIb V-8	
Wing:	Area	17.0 m²

Had the V37 design progressed, it would certainly have been fitted with the standard fighter armament of two synchronized machine guns. However, the V37 arrived too late to be evaluated, let alone produced in quantity, so its performance and flying qualities are not known. Even its dimensions are unknown.

Fokker V 37

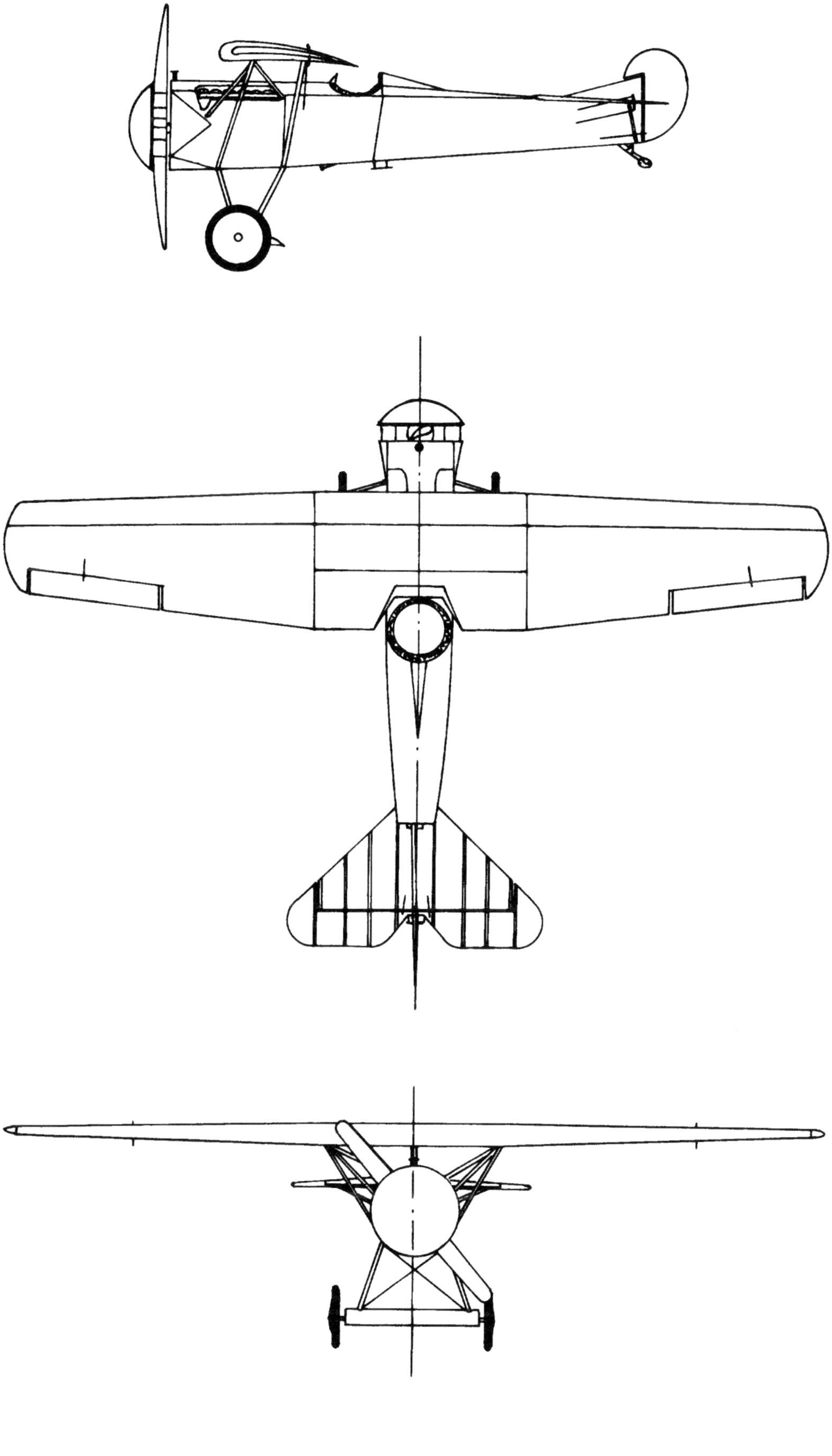

Above: The Fokker V37 armored fighter prototype combined the configuration of the earlier Fokker V27 with a distinctive propeller spinner and the 195 hp Benz Bz.IIIb V-8 water-cooled engine. (Peter M. Grosz Collection/STDB)

Below: The entire forward fuselage of the Fokker V37 was armored with 2.5mm steel to protect the pilot, engine, and radiator from rifle-caliber ground fire. The extent of the armor along the cockpit is clearly visible. The fan blades that forced cooling air through the radiator. (Peter M. Grosz Collection/STDB)

Above: The Fokker V37 armored fighter prototype uncovered fuselage showing its armor plate and steel-tube rear fuselage. The engine installation and radiator fan blades are visible. (Peter M. Grosz Collection/STDB)

Below: Rear quarter view of the Fokker V37 showing its armored engine and cockpit coupled with the steel-tube fuselage and tail structure common to Fokker aircraft. (Peter M. Grosz Collection/STDB)

Above & Below: Forward and rear views of the Fokker V37 armored fighter prototype emphasized how streamlined it was despite the armored plate forward fuselage. (Peter M. Grosz Collection/STDB)

Above: Rear quarter view of the Fokker V37 shows its headrest fairing the armored plate behind the pilot. (Peter M. Grosz Collection/STDB)

Fokker V38

Above & Below: The Fokker V38 was an enlarged, two-seat derivative of the BMW-powered D.VII. It was the prototype of the Fokker C.I that was produced in series postwar. (Below Peter M. Grosz Collection/STDB)

The V38, W/N 3658, was an enlarged two-seat derivative of the D.VII with the 185 hp BMW IIIa. It was the prototype of the Fokker C.I, which was only built in series after the war. Armament was one synchronized gun for the pilot and a flexible gun for the observer.

Fokker V 38

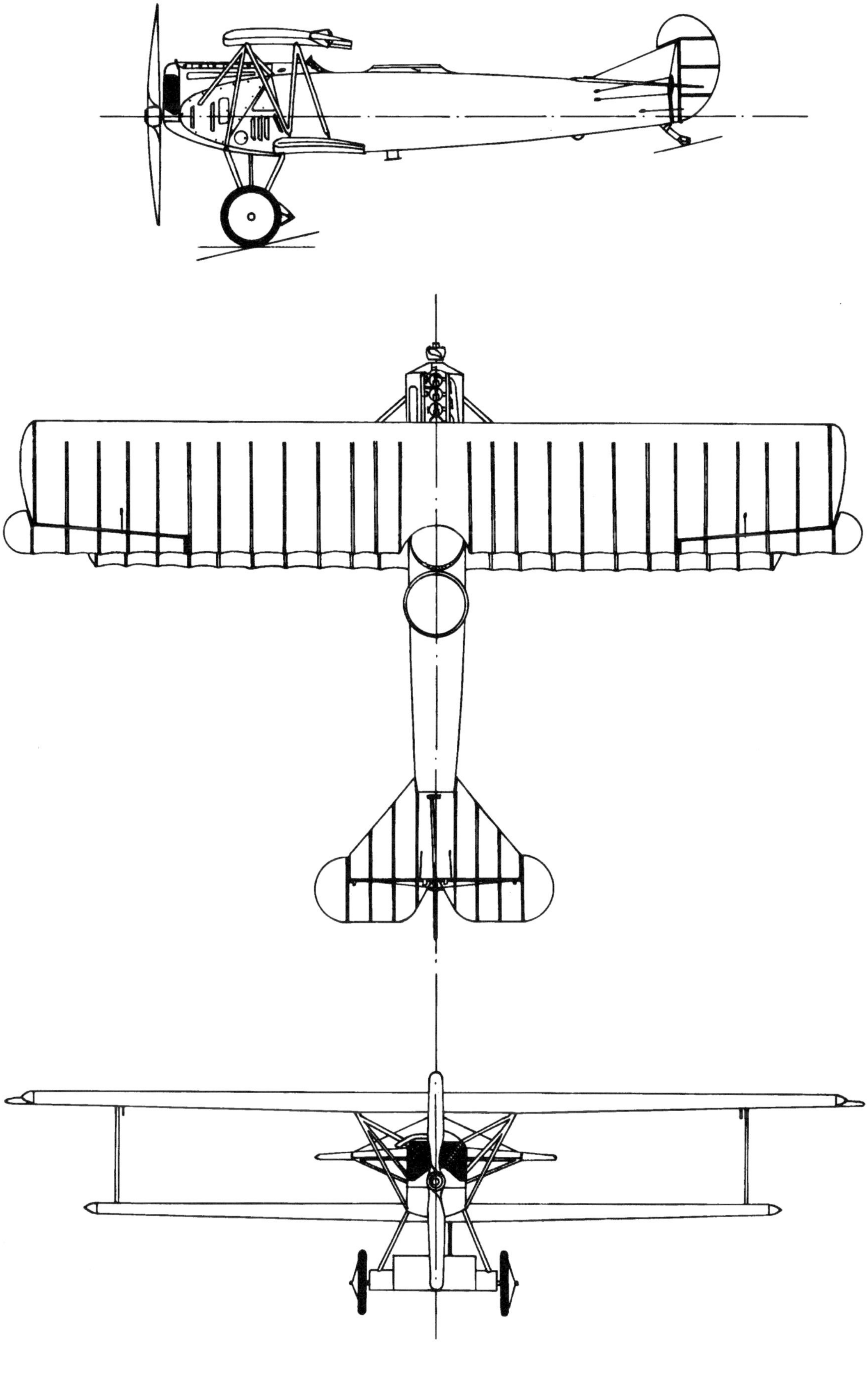

Above: The Fokker V38 was an enlarged, two-seat derivative of the BMW-powered D.VII. It was the prototype of the Fokker C.I that was produced in series postwar. (Peter M. Grosz Collection/STDB)

Above: The uncovered fuselage and tail of the Fokker V38 prototype of the Fokker C.I showing its steel-tube structure that was standard for Fokker.

Above & Below: The Fokker V38 was an enlarged, two-seat derivative of the BMW-powered D.VII. It was the prototype of the Fokker C.I that was produced in series in Holland postwar. The longer wing span compared to the D.VII is evident. (Peter M. Grosz Collection/STDB)

Fokker V39

Above & Below: The Fokker V39 was a postwar sports aircraft. (Peter M. Grosz Collection/STDB)

The Fokker V39 was a sports plane offered postwar. It was very similar to a scaled down D.VIII powered by a 50 hp Gnome rotary.

Fokker V39 Specifications

Engine:	50 hp Gnôme	
Wing:	Span	7.0 m
	Area	10.2 m^2
General:	Length	5.02 m
	Height	2.50 m
	Empty Weight	210 kg
	Loaded Weight	415 kg
Climb:	1000m	6 min
	2000m	17.5 min

Fokker V 39

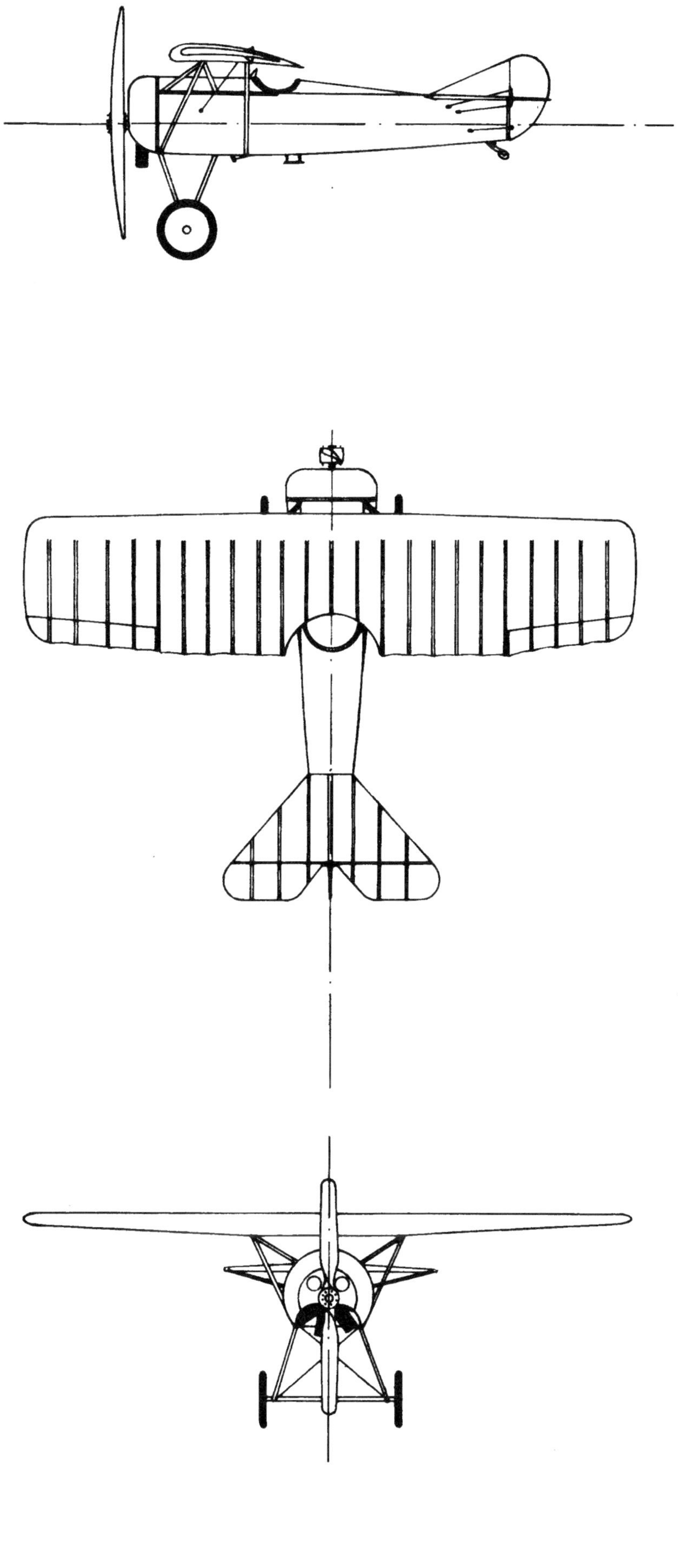

Above, Below, & Facing Page: The Fokker V39 in Dutch insignia postwar. Its resemblence to the D.VIII is clear. (Peter M. Grosz Collection/STDB)

Fokker V40

Above & Below: The Fokker V39 was a postwar sports aircraft.

There are two contradictory specifications for this type designation. According to press reports from the first postwar years and the Fokker type list of the time, the V40 was a small, single-seat sporting cockpit with a 30–35 hp Anzani radial engine. This is the aircraft shown here. It is said to have reached speeds of up to 120 km/h, although the engine caused strong vibrations. This aircraft was built in Schwerin in April 1919.

On the other hand, according to contract 344-T of June 30, 1920, Fokker was to supply the American War Department with two "V40" single-seat monoplanes as cantilever parasol aircraft, equipped with 300 hp Wright-Hispano engines. In America, the aircraft were called F VI or PW-5.

Fokker V40 Specifications

Engine:	35 hp Anzani	
Wing:	Span	5.9 m
	Area	7.0 m^2
General:	Length	3.94 m
	Empty Weight	160 kg
	Loaded Weight	266 kg
Maximum Speed (near ground level):		111 km/h

Fokker V 40

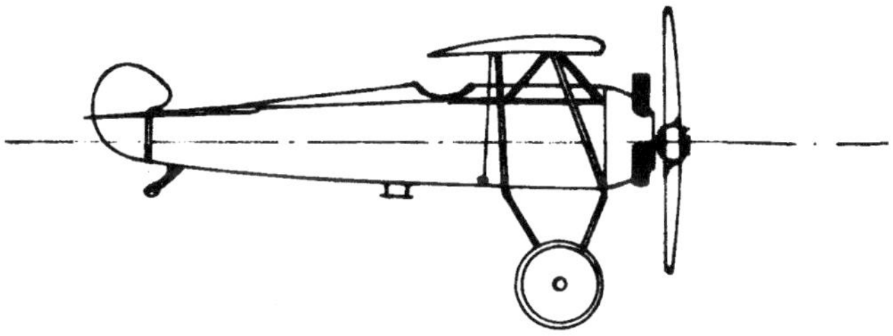

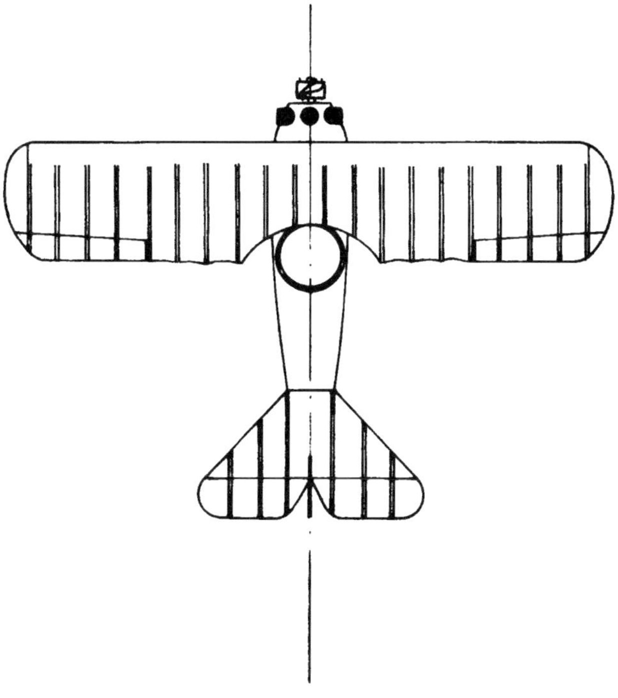

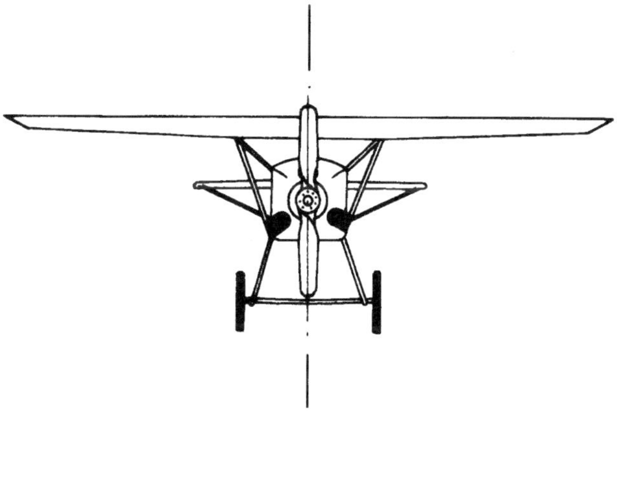

This Page & Facing Page, Bottom: The Fokker V40 under construction. (Peter M. Grosz Collection/STDB)

Above: The uncovered Fokker V40 airframe in the foreground. The V37 prototype airfame is in the background and provides a good size comparison. (Peter M. Grosz Collection/STDB)

Fokker V41

Above: This two-seater variant of the Fokker D.VII has a modified raised fuselage contour and also the cut-out at the trailing edge of the wing is smaller than on the D.VII 10415/18 in the other pictures on the spread. (Peter M. Grosz Collection/STDB)

Fokker V41 Specifications

Engine:	185 hp BMW IIIa	
Wing:	Area	25.2 m^2
General:	Empty Weight	723 kg
	Loaded Weight	1300 kg

According to the Fokker type list drawn up at the end of the war, the V41 was a two-seat biplane with N-struts and a 185 BMW IIIa engine. So the V41 was likely a designation for one of the two-seater versions of the D.VII known from photos. Various enlarged derivatives of the D.VII were investigated and tested. On May 31, 1919, the pilot Parge in a Fokker D.VII with a large wing, as later used in the C.I, reached an altitude of 8000 meters in only 36 minutes. Likewise, there have been different two-seat school and travel variants of the D.VII. Whether these different experimental aircraft also carried a V-number is unknown to this day.

Left: Ernst Udet in the unarmed Fokker 10415/18 postwar.

Above: The military serial number of this two-seat D.VII is 10415/18 – without any reference to the special version. The upper wing cutout is larger than the aircraft on the facing page. It is not V41 but is being flown by Ernst Udet at Kbely airfield, Prague in 1919. (Peter M. Grosz Collection/STDB)

Above: The Fokker 10415/18 was derived from the Fokker D.VII with a 185 hp BMW IIIa . Here it is being flown by Ernst Udet postwar. This photo also shows the fuel line between the undercarriage tank and the fuselage. This aircraft may be one of the many conversions to D.VIIs to two-seaters, many undocumented. The undercarriage fuel tank clearly indicates it is not a two-seat conversion of a standard D.VII.

Fokker V42

Above: The V42 was quickly placed on a shallow float. The cartoons on the nose of the glider show that it is one and the same aircraft as in the land version shown below.

Below: This small tow glider was developed in Schwerin as early as 1918, but was first tested in Holland. Shown here is the land version equipped with two wheels.

V42 was an unpowered, should-wing glider of mixed construction with interchangeable wheel undercarriage or a wide float without step placed under the fuselage. The type designation is given by Hegner; data and development history were published in 1943. According to Reinhold Platz's recollection, in 1917 Fokker allegedly wanted to have a D.VIII without engine towed on a rope behind a converted two-seater D.VII equipped with tail protection. Platz built a special design of the glider, which was completed in Schwerin in 1918 but not tested there because of the end of the war. The fuselage and tail unit were welded from steel tubing and fabric-covered, with the leading edge extended to the trailing edge of the wing. The wooden wing had a plywood nose and fabric covering. The aircraft was taken to Holland, where the first experimental tow behind a motorboat on the Ymuiden-Amsterdam canal took place in about mid-1919 or 1920, piloted by Adolf Prag. Previously, the glider had been towed unmanned with ballast.

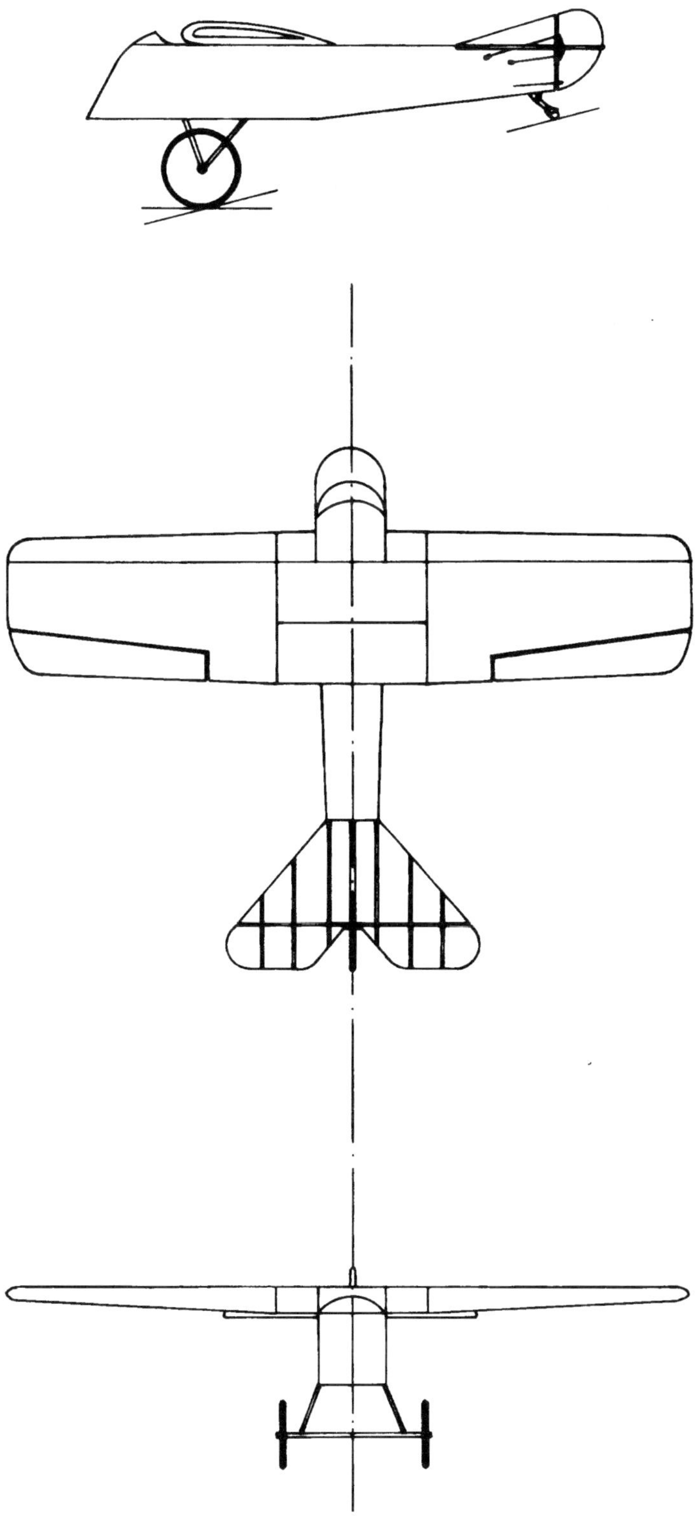

Fokker V43

Above: The Fokker V43 was built in Schwerin in 1919 as a prototype trainer with side-by-side seating for instructors and students.

This prototype of a high-wing training aircraft was built in Schwerin in 1919. Instructor and student sat side by side in the plane. Later, in the Netherlands, the prototype derived from this was designated the S I. While the prototype had a continuous trailing edge with correspondingly poor upward visibility, the S I was given a crude wing cutout. It is unknown how many aircraft were built in total. The US Army Air Service purchased an S I in 1922 and designated it a TW4. Two S Is were delivered to Soviet Russia. The V43 remained in service until 1921, when it fell victim to a landing accident in Amsterdam in October.

Fokker V43 Specifications

Engine:	75 hp Mercedes	
Wing:	Span	12.6 m
	Area	26.5 m^2
General:	Length	8.26 m
	Height	2.56 m
	Empty Weight	609 kg
	Loaded Weight	883 kg
Maximum Speed:		140 km/h
Landing Speed:		65 km/h

Fokker S I Specifications

Engine:	90 hp Curtiss	
Wing:	Span	12.74 m
General:	Length	8.53 m
	Height	2.59 m

Left: The Fokker V43 in front of the Schwerin hangar. A door was built into the side of the fuselage to enable access to the aircraft's cockpit.

Fokker V 43

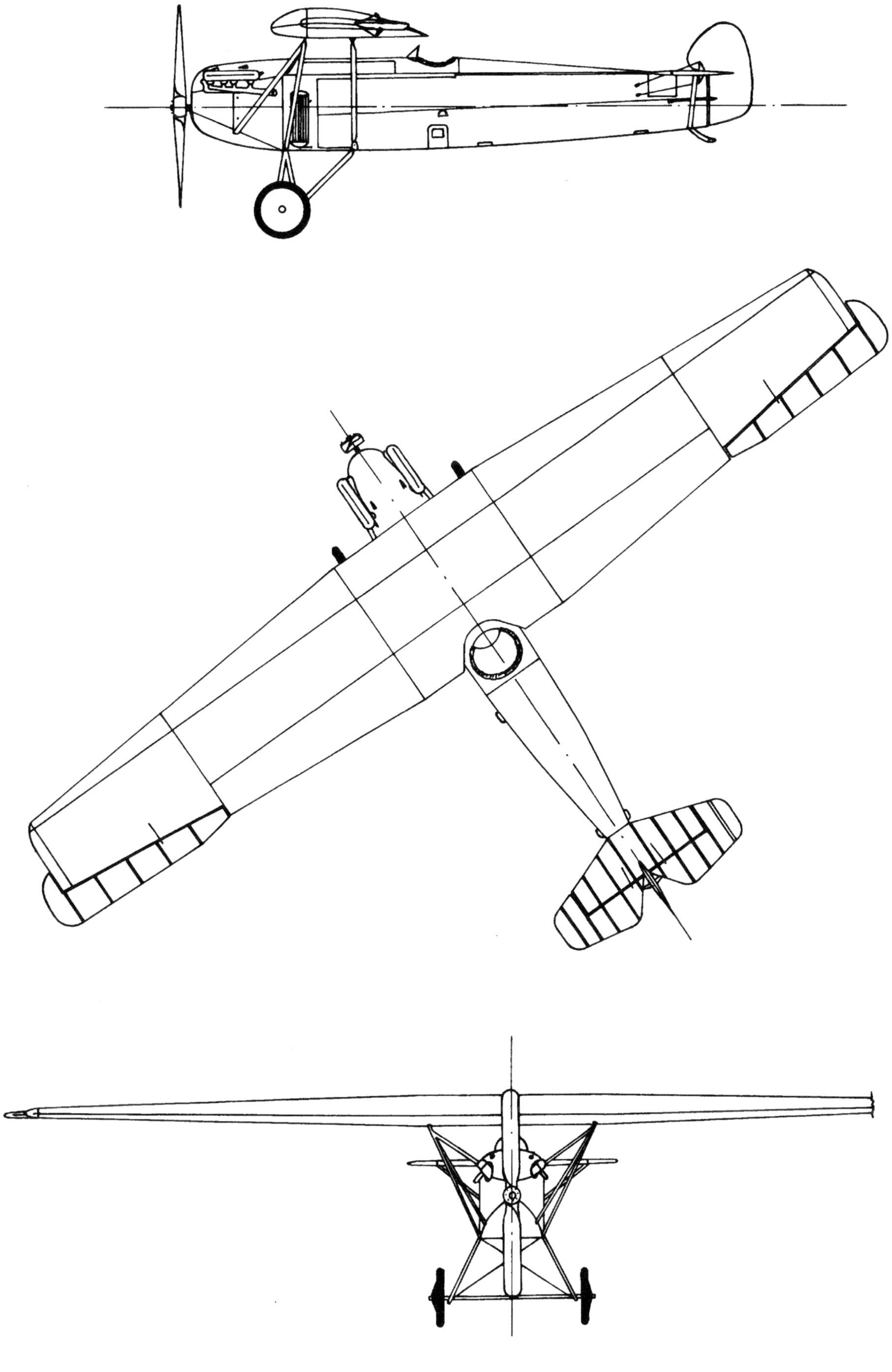

Fokker V44 & V45

Above & Below: The V45 (Works Number 1500) was given the German civil registration issues in May 1920. The V45, powered by a 185 hp BMW IIIa , was the prototype commercial aircraft from which the Fokker F II production aircraft was derived. In contemporary documents and publications, the V45 is also referred to as 'V1' or 'F I', where the letter 'V' here stands for 'commercial aircraft'. The crew was one or two pilots with four or five passengers. (Peter M. Grosz Collection/STDB)

Fokker V45 Specifications

Engine:	185 hp BMW IIIa	
Wing:	Span	17.24 m
	Area	42.0 m^2
General:	Length	11.63 m
	Height	3.67 m
	Empty Weight	1200 kg
	Loaded Weight	1900 kg
Maximum Speed:		150 km/h
Endurance:		7 hrs

Above: For the transfer to the Netherlands, the V45 was given a modified livery and the lettering 'Fokker.' The V40 is at left. (Peter M. Grosz Collection/STDB)

The Fokker V44 was essentially an enlarged V43 powered by a 185 hp BMW IIIa having a wing span of 20 m and length of 13 m. The V45 (work number 1500) was a more advanced aircraft, also powered by a 185 hp BMW IIIa, with enclosed cabin for the pilot and passengers. V45 was slightly smaller and more streamlined than the V44 and was the prototype of the production Fokker F.II transport.

Above: The parasol-designed glider described in the literature as V30, which is said to have originally been towed to the target by a D.VII and filled with explosives to have crashed as a "glide bomb" after separation from the tow plane, cannot have carried this V number because of the documented records to the contrary. It is said to have been created from the V26 by removing the engine and was exhibited at the Paris Aerosalon in 1921. Fokker filed a patent for the towing device in his name in the USA on June 16, 1921, and received it on June 6, 1922, under the number 478.084. There is no contemporary confirmation of the type designation of this glider.

Fokker D.VI

Above: A very early-production Fokker D.VI on the airfield at Schwerin-Görries. The cowling is still unpainted, the machine guns remain to be fitted, and the first version of the *Balkenkreuz* marking with the 15 cm white borders is applied on wings and fuselage. (Reinhard Zankl)

Between 20 January and 12 February *Idflieg* held the First Fighter Competition at Berlin-Adlershof, and Fokker demonstrated both inline-engined and rotary-engined prototypes there. Entered by Fokker in the latter category were three biplanes, the V 9, powered by the 110 hp Oberursel Ur.II, the V 13-I, powered by the eleven-cylinder Oberursel Ur.III, which was rated at 145 hp, and the V13-II, which was powered by the 160 hp eleven-cylinder Siemens & Halske Sh.III engine. Both eleven-cylinder engines were still considered experimental, but were hoped to be ready for mass-production soon.

The Siemens-Halske Sh.III was demonstrated in several other types by other manufacturers, too, and turned out to be troublesome in most of these at the competition. Yet, it entered production before the Ur.III did, although its initial service would quickly prove the need for further perfection to the engine.

While being of very similar appearance, the two V 13 prototypes were 33 cm longer than the V 9, while having 5 cm shorter wingspan.

Of the three engines mentioned, only the Ur.II was already in mass production and frontline use as the powerplant of the Fokker Dr.I triplane. The Fokker Dr.I was introduced into widespread combat use with several elite *Jagdstaffeln* around the turn of 1917/18. Pilots very much appreciated the superb rate of climb and excellent maneuverability offered by it. However, it quickly became obvious to the men who flew the type that its horizontal speed was insufficient to keep up with many of the enemy fighters it faced in daily combat.

The triplane design, while offering excellent climb and maneuverability, simply produced too much drag. Since the 110 hp Oberursel Ur.II that powered the Dr.I was in full-scale production, and its more powerful successor, the eleven-cylinder Ur.III, promising to deliver around 145 hp, was still not ready to enter service, a new airframe providing faster

Above & Below: The first production D.VI 1631/18, the true production prototype, was photographed at Adlershof during its *Typenprüfung* in April/early May 1918. Its works number was 2613. The D.VI had excellent visibility for the pilot, was very maneuverable, and was light and sensitive on the controls, a real pilot's airplane. It just needed more power than its 110 hp Oberursel UR.II offered to be a first-class fighter. Comparison of the D.VI specifications with the V17 monoplane prototype using the same engine reveals the great improvement in performance of the monoplane configuration. The V17 probably should have been selected for production instead of the D.VI because its performance was far superior, but its monoplane configuration was suspect and the wing obscured the pilot's downward field of view.

Above: The tail of Fokker D.VI 1631/18 has been put up on a trestle in order to allow a frontal shot to be taken. (Peter M. Grosz Collection/STDB)

speed with the existing engine, had to be designed.

The V 9 and the two V 13s had performed well enough at the competition for *Idflieg* to award Fokker a production order for 120 airframes that were officially designated D.VI in March. These were to be powered by the readily available Oberursel Ur.II, so the smaller V 9 served as the base version for the final production version of the D.VI.

At this point two important factors have to taken into account. First, Fokker Triplane production at Schwerin was rapidly approaching the end. Out of 320 examples that had been ordered in total, only 21 machines remained to be completed after 31 March. With production of the triplane tapering off, a faster rotary-engined successor was needed. Since the D.VI offered better speed and slightly better low-altitude climb than the triplane, it was chosen as the follow-up model in the Schwerin production halls.

Second, at the same time, another Fokker type had emerged as the winner in the inline-engined category, the Fokker D.VII. This is the type that was truly intended as a replacement for both the outdated Albatros D.V/Va as well as the Fokker Dr.I, as is graphically illustrated by the initial production orders for the D.VII. Right after the end of the competition, in February, 400 D.VIIs were ordered from Albatros, while an initial order for 300 examples was awarded to Fokker. For the Fokker company, this meant that while the Dr.I was being phased out of production, preparations had to be made to begin the manufacture of two different airframe types simultaneously. The fact that these were powered by different engine types complicated matters further.

Fokker took a very economical approach to face this challenge. A fuselage very closely based on that of the Dr.I was chosen for the D.VI production version. Length of both fuselages was practically identical (5776 mm for the D.VI versus 5770 mm for the Dr.I). Of course, the fuselage between the cockpit and the engine was modified for the biplane configuration; the attachment for the middle wing was eliminated and the lower wing cutout was enlarged because it had a greater chord. The Dr.I tailplane layout was carried over to the D.VII, too.

For the production wings of the D.VI an equally clever solution was found. The chord of the lower wings for both the D.VI and the D.VII was identical (1.20 meters), and the same was true for the upper wings of both types (1.60 meters). Consequently, identical wing ribs could be manufactured for both. In fact, when comparing drawings of the wings of the D.VI to those of the wings of the Fokker D.VII, it becomes apparent that the D.VI wing design was nothing but a shortened D.VII wing. The wing ribs were of identical proportions on both types, and even the ailerons were alike.

These similarities in design considerably eased serial production of the D.VI, and the first example of the new type (D.1631/18) was accepted at Schwerin on 26 April 1918.

From a logical standpoint, it seems that the D.VI was primarily put into production to make the best use of the UR.II engines as they were coming off the production line. It is likely that Fokker was requested to manufacture airframes

Above: Fokker D.VI 1689/18 is seen here at the *Flugpark* in the area of *Armee-Abteilung* B. Numerically, this was the last aircraft built, being accepted on 8 June, and photographed here on 21 June. However, chronologically the last two D.VIs to be accepted at Schwerin for the *Fliegertruppe* were 1637/18 and 1639/18 on 1 August 1918.

at roughly the same rate as he received the rotary engines that powered it, in order not to interfere with Fokker D.VII production, which had absolute priority. This, however, is a mere assumption that is not backed up by available contemporary references.

An initial order for 120 machines was placed, and this low number clearly indicates that the Fokker D.VI was not intended to serve at the front in large quantities. Eventually, as the development of better-performing types progressed, even this relatively modest production order would be cut in half.

The first 23 production examples of the D.VI were accepted at Schwerin in May 1918, and a further 26 followed in June, none in July, and the last nine in August. After 60 aircraft were completed, Idflieg terminated half of their production order in favor of the Fokker E.V.

Both types were powered by the same engine, but the E.V was faster and had a better rate of climb. Being a parasol monoplane, it also offered better downward visibility than a biplane, and the combination of these three factors spelled the end for the D.VI. Additionally, in all probability it was also easier to manufacture than all previous Fokker types. Thus, production of the D.VI turned out to be very short-lived, making room for the E.V as soon as it passed the customary static loads test.

By the end of June, 21 Fokker D.VI were listed in the front line inventory, this number modestly increasing to 27 at the end of August. Compared to the 407 Fokker D.VIIs already at the front and 118 Fokker Dr.Is still serving with various units on 30 June, it is readily apparent that the D.VI was not intended to play a major role in frontline combat.

This is proven by the fact that no *Jasta* which had been previously equipped with the Fokker Dr.I received the D.VI as a replacement. This role was fulfilled by the Fokker D.VII, which had become the favorite mount of the Jasta pilots in a very short time. Instead, The D.VI was supplied to units serving on "quieter" sections of the front, and to home defense units.

Royal Bavarian *Jagdstaffel* 80 was one of the former units. At least six examples were supplied to the *Jasta* in June 1918 to partially replace their older Albatros D.Va fighters. The pilots were pleased with their new planes, which offered good maneuverability and immediate responsiveness to the controls.

Few reports of pilots having flown the D.VI are known, but they generally praised the type. However, overall performance of the D.VI was not as good as that of the Fokker D.VII, especially at altitude. The low-powered nine-cylinder rotary engine lost performance at altitude, and since combats took place at ever increasing altitudes, this was a considerable penalty.

Small numbers of the D.VI are known to have been delivered, amongst other units, to *Jastas* 64, 71, and 75, as well as to several *Kests*. The fast rate of climb of the rotary-powered fighter made them an ideal candidate for the interceptor role, and the engine required almost no warming up, unlike inline engines.

The *Jagdstaffelschulen* also received several examples, but the low number of 60 aircraft produced prevented the D.VI from making a lasting impression, either on the German pilots or their opponents. However, it is interesting to note that no example is known to have been captured intact by the Allies before the Armistice.

As production of the Fokker D.VI was getting underway, the first examples of a more powerful rotary engine became available. The nine-cylinder 170 hp Goebel Goe.III engine had successfully completed the prescribed 50-hours duration test in April, and twelve D.VI (military numbers 1637/18, 1639/18 & 1640/18, 1643/18 and 1679/18–1676/18) received the new engine. The Goebel had a slightly bigger diameter (1050 mm) than the Ur.II (960 mm), but was also a nine-cylinder engine. Apparently *Idflieg* intended to use the D.VI as a frontline evaluation mount for the Goebel, which actually delivered between 165–178 hp depending upon altitude. In theory, the increased performance promised to solve the speed issues while giving an even faster rate of climb.

Unfortunately, no records or reports of this evaluation have been found, but the facts speak for themselves. No examples of the Goebel engine were later seen in the Fokker E.V when it reached the front in August, and it only reappears later in October 1918 in several prototypes during the Third Fighter Competition in Adlershof. Just before the Armistice, it was intended as the standard powerplant for the Kondor E.IIIa parasol fighter, but by then it was too late.

Fokker D.VI Specifications

Engine:	110 hp Oberursel Ur.II	
Wing:	Span upper	7.20 m
	Span lower	5.81 m
	Chord upper	1.60 m
	Chord upper	1.20 m
	Gap	1.25 m
	Area	17.1 m^2
General:	Length	5.78 m
	Height	2.65 m
	Empty Weight	395 kg
	Loaded Weight	588 kg
Climb:	1000 m	2.6 min.
	2000 m	5.9 min.
	3000 m	10.7 min.
	4000 m	17.3 min.
	5000 m	28.0 min

(The Kondor E.III and E.IIIa are documented in *German Aircraft of Minor Manufacturers in WWI,* volume 1.)

The fact that no photograph showing a D.VI powered by the Goe.III serving with a *Jasta* or *Kest* is known is also noteworthy. Obviously, the engine was not yet ready for frontline service and required further fine-tuning, just as was initially the case with the Siemens-Halske Sh.III.

In retrospect, the main purpose of the D.VI was to provide an interim solution that would be an improvement over the triplane and that could make good use of the available Oberursel engines. But the appearance of the simpler, better-performing Fokker E.V parasol quickly ended its short career. Yet its basic design would resurface once more in the shape of the Fokker V.33 just before the end of the war.

Fokker D.VI Production Orders

Order Date	Qty	Serial Nos.	Notes
March 1918	120	D.1630–1749/18	60 aircraft, serials 1630–1689/18, were built. The rest were cancelled to produce the D.VIII that used the same engine.
May 1918	150	Unknown	Cancelled to produce the D.VIII.

"The Fokker D.VI aircraft flown by the *Staffel* during the month of August experienced frequent cylinder seizures of the Le Rhône 110 hp rotary engine. The Fokker D.VI was very maneuverable at low altitudes, but it took about one hour to reach the altitude of 5000 meters. It was very slow and unmaneuverable at these altitudes. In Fokker D.VI 1679/18, one cylinder had seized, because the pilot did not immediately recognize the error and continued to fly at full throttle, tearing the connecting rod from the pistons and pushing the broken connecting rod between the threads of the cylinder and the casing, causing the cylinder to break loose and break the main spar of the lower wing. The aircraft crashed out of control [*steuerlos*] from about 1000 meters. The Ricinus substitute oil No. B237 [and] B238 has served well as long as the *Staffel* has flown the Fokker D VI."

Source: The Activity Report of the Württbg. *Jagdstaffel* Nr. 64 for the period from August 1 to September 30, 1918.

Above: A *Kest* 1a Fokker D.VI on the airfield at Mannheim. The cowling has been painted in a light color interpreted as light blue for the profile. Personal marking of a double pennant, interpreted as black and white, is seen below the cockpit. The fuselage cross has been converted from the earlier full white border version by extending the black cross arms. For some reason there are contradicting overall dimensions given for the D.VI in various contemporary sources. (Peter M. Grosz Collection/STDB; also Bruno Schmäling).

Fokker D.VI of *Kest* 1a.

Above & Below: A single Fokker D.VI of *Kest* 1b is seen here with several Fokker D.VIIs at Karlsruhe. The fuselage and rudder cross have been converted by extending the black cross arms. The Fokker company logo decal is visible on the bottom of the rudder, and next to it the Fokker Werknummer has been marked. Unfortunately, it is too indistinct to be read clearly. (Above: Greg VanWyngarden, Below: Reinhard Zankl)

Above: A view of the same aircraft at Karlsruhe. Just visible is the lower wing cross, the vertical cross arm extends from leading to the trailing edge of the wing. The tonality of the two-color fuselage band looks a bit different in all four photos of the plane, and it has been interpreted as yellow-red-yellow for the profile. Karlsruhe is located in the Grand Duchy of Baden, and these are the colors of the Baden coat of arms. (Reinhard Zankl)

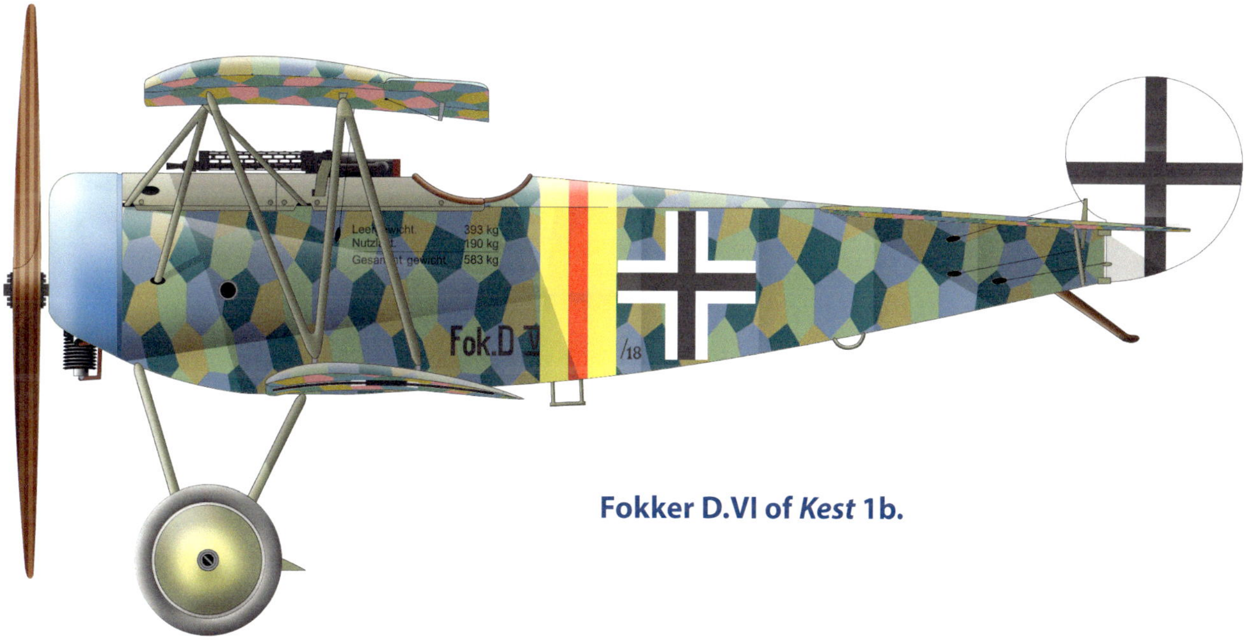

Fokker D.VI of *Kest* 1b.

Above: While this photo of the *Kest* 1b D.VI is of less than perfect quality, it shows the small "thin" *Balkenkreuz* on the upper wing to good advantage. None of the available photos of this aircraft shows the military or factory works number clearly. (Peter M. Grosz Collection/STDB)

Above: Fokker D.VI 1688/18 is seen here on the *Kest* 4b airfield at Freiburg. The aircraft is still in factory finish, seen behind it is Albatros D.Va D.5629/17 "4". (Reinhard Zankl)

Above: Groundcrew members of *Kest* 4b assemble in front of the new plane for a group photo. This may have been the first fighter powered by the Oberursel Ur.II engine delivered to the unit, and the mechanics were probably curious to have a look at the engine. (Peter M. Grosz Collection/STDB)

Above: Full view of the same D.VI minus cowling; note the full-chord *Balkenkreuz* on the lower wing having 5:4 proportions. A lineup of *Kest* 4b Albatros D.Va fighters serves as the backdrop, with D.5351/17 "3" seen at right. (Peter M. Grosz Collection/STDB)

Above: A Fokker D.VI serving with *Kest* 7 at Düsseldorf with late-war think national markings. The pilot has been identified as *Lt*. Hoffmann. (Peter M. Grosz Collection/STDB)

Above: Fokker D.VI taken over by the USAS postwar as part of the Armistice agreement. Three D.VI fighters, 1639/18, 1655/18, and 1676/18, were shipped to the US from Romorantin on 2 April 1919. At least one was assigned USAS serial number AS 94033 and McCook Field number P-385. The tail markings suggest possible previous service with *Jasta* 71. (Peter M. Grosz Collection/STDB)

Above: A pilot and seven members of the maintenance crew of *Jasta* 75 pose in front of their Fokker D.VI at Habsheim airfield. The hangar marked "A10" behind them was formerly used by the Aviatik company, the inscription still being faintly visible on the roof arch. (Peter M. Grosz Collection/STDB)

Below: Full view of the same aircraft, showing that the bottom of the axle wing was painted light blue. On the lower part of the cowling, the dark green paint has been worn off by the castor oil spewing out of the Oberursel engine. (Peter M. Grosz Collection/STDB).

Above: *Jasta* 80b *Staffelführer Oblt*. Erwin Wenig with his dog in front of his Fokker D VI 1661/18; the Fokker *Werknummer* 2643 is visible on the struts. His personal marking consisted of a black fuselage section aft of the cockpit, edged by narrow black bands on both sides. (Bruno Schmäling)

Below: Frontal view of Wenig´s new Fokker taken around the same time as the photo above. The streamlined rear-view mirror in the upper wing cutout and the offset "Axial" decals help to identify the aircraft in other photos. (Bruno Schmäling)

Fokker D.VI 1661/18 of *Oblt*. Erwin Wenig,
Staffelführer of *Jagdstaffel* 80, Summer 1918

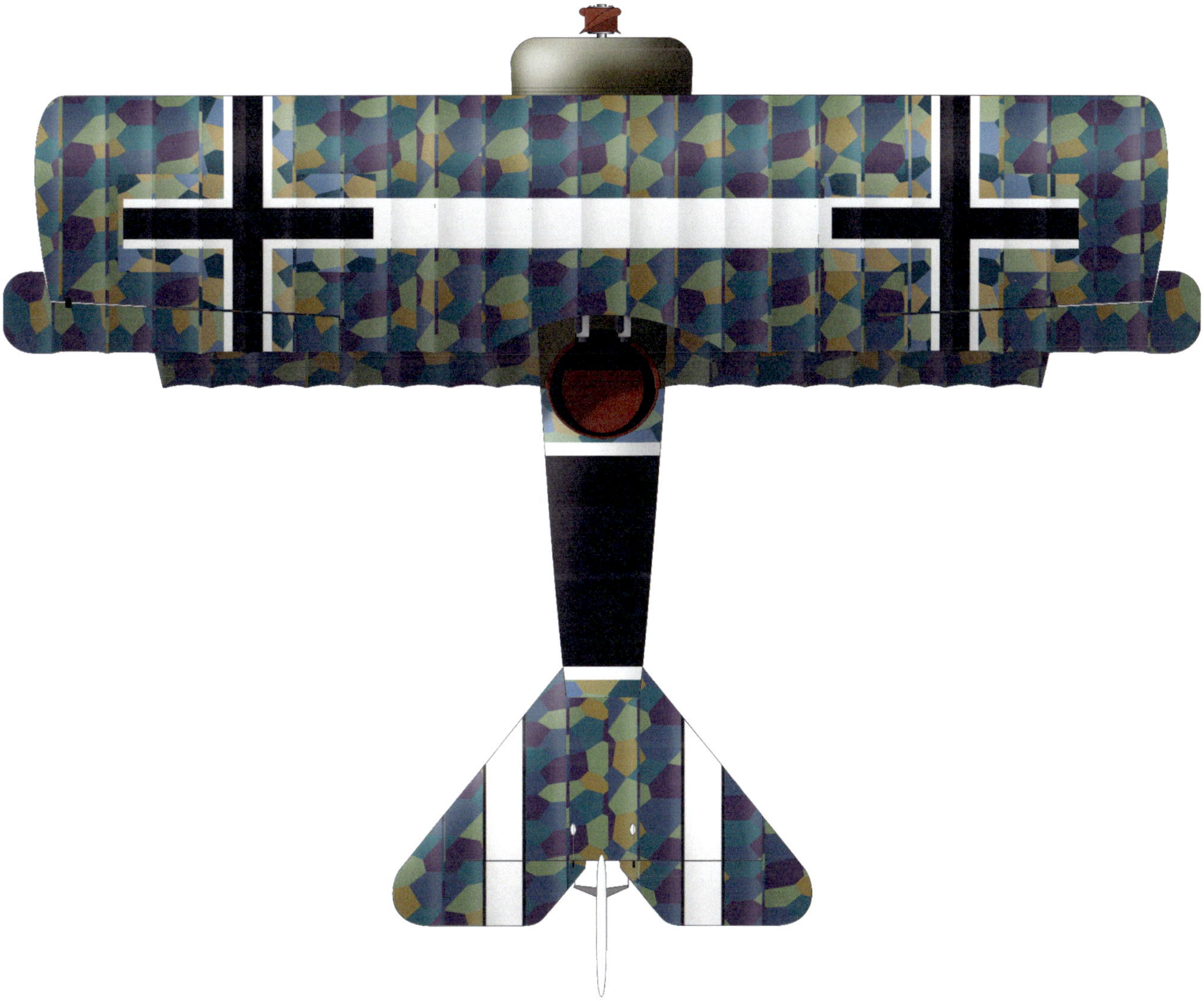

Above & Below: Wenig´s Fokker D.VI 1661/18 shows up in several *Jasta* 80b photos, and seen here are two members of the *Jagdstaffel* posing in front of this plane. In the lower photo, the units special duties officer (not pilot) *Oblt.* Max Speidel seized the opportunity for a snapshot. (Peter M. Grosz Collection/STDB; above also Bruno Schmäling).

A rare lineup of Fokker D.VI fighters of *Jagdstaffel* 80b at Morsberg (now Marimont-lès-Bénestroff), with Wenig´s 1661/18 heading the six planes. The photo was taken at the same time as the two portrait shots on the opposite page. (Alex Imrie)

Above: Fokker D.VI 1661/18, flown by *Oblt*. Erwin Wenig shows the *Jagdstaffel* 80 unit marking of a white stripe, applied here with a narrow black border, on each side of the elevator. A white stripe was painted on the center section of the upper part of the top wing as *Staffelführer* identification marking. (Bruno Schmäling)

Above: At some later point, *Jasta* 80b member *Lt.d.R.* Josef Filbig was photographed with the aircraft. By then, the black center section had been re-painted Bavarian blue, and the name "*Jette*" had been added below the cockpit. (Tobias Weber)

Below: Photo retouching 1918-style! While all details of the uniform in this photo match those of Josef Filbig seen in the photo above, the head of another unknown individual has been montaged into the photo here.

Fokker D.VI 1661/18
***Lt.* Josef Filbig**
***Jasta* 80b, Summer 1918**

Note: Originally this aircraft was flown by *Oblt*. Wenig. When *Lt*. Filbig joined the *Jasta* he took over 1661/18 and painted the black fuselage band blue.

Below: The prop and rear-view mirror indicate that this photo again shows 1661/18 in yet another, later paint scheme. The cowling now appears in a light color, while the white borders of the mid-fuselage marking have disappeared. This section now looks dark again. While difficult to make out, the facial features of the pilot do resemble those of Filbig. (Peter M. Grosz Collection/STDB)

Lt. Kurt Seit and his Fokker D.VI *Fratz* at *Jasta* 80b. Bavarian colors (blue & white) were used for the personal markings. The aircraft carried 5-color printed camouflage fabric on its wings. (Peter M. Grosz Collection/STDB)

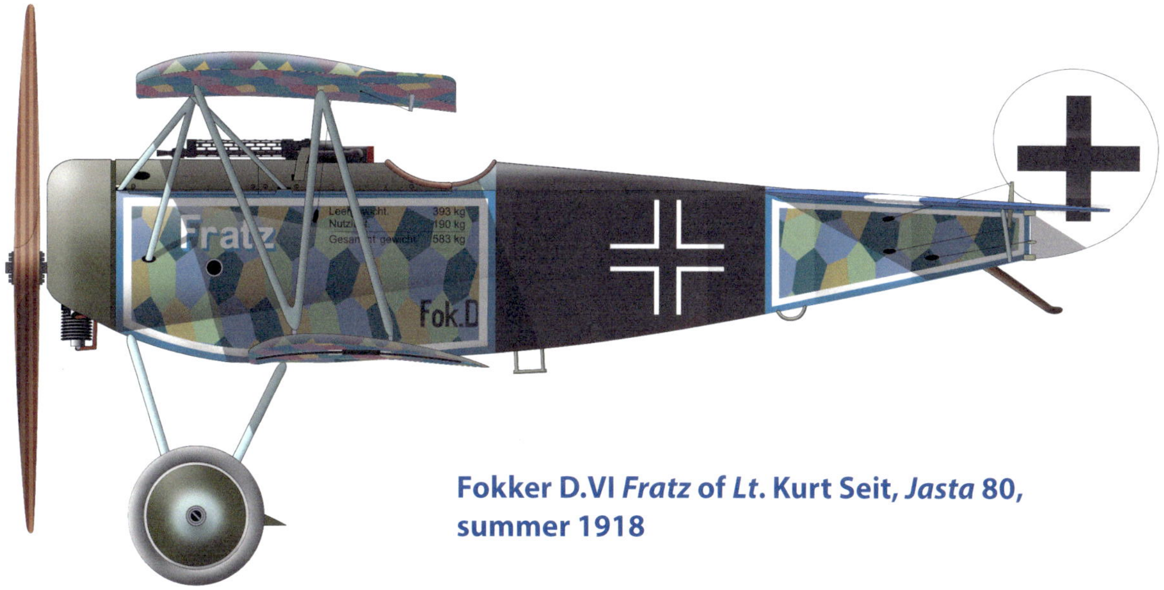

Fokker D.VI *Fratz* of *Lt*. Kurt Seit, *Jasta* 80, summer 1918

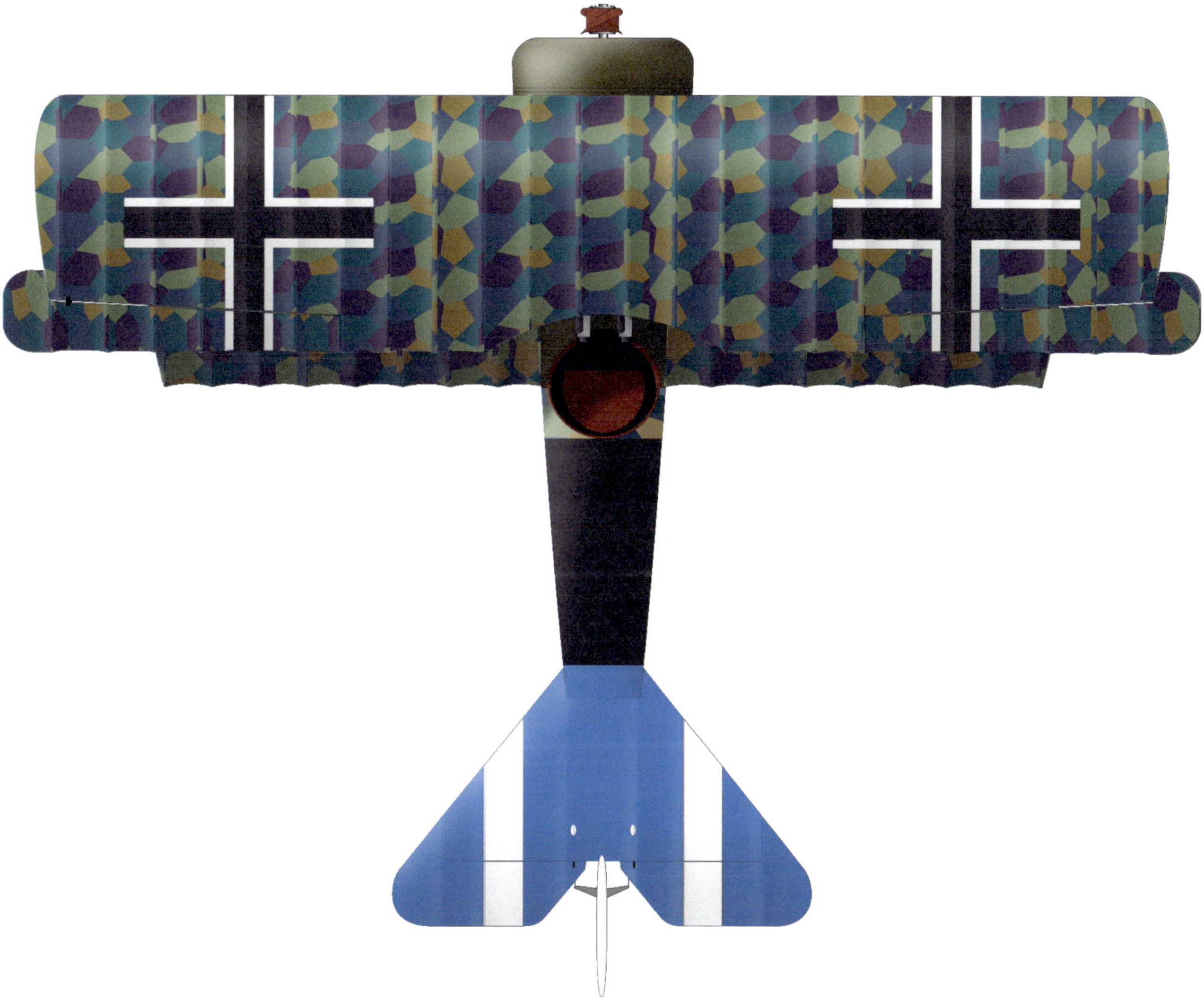

Above: *Lt*. Otto Fuchs of *Jasta* 77b photographed in the cockpit of a Fokker D.VI of *Jasta* 80b while he visited the unit. (Dr. Volker Koos)

Facing Page: A better look at the face marking of "Fratz" is provided by this snapshot of two mechanics. (Bruno Schmäling)

Right: Pilot Otto Fuchs in a Fokker D.VI of *Jasta* 80b. At the time he was serving in *Jasta* 77b, and *Jasta* 77b shared an airfield with *Jasta* 80, where both were part of "*Jagdgruppe* Raben" which consisted of *Jastas* 3, 18, 54, 77, and 80. Otto Fuchs told historian Bruno Schmäling that he and other pilots of *Jasta* 77 visited *Jasta* 80 and in reverse. When he was at Morsberg he of course flew the Fokker D VI, but he was not very impressed. (Dr. Volker Koos)

Above & Below: These photos may show the same Fokker D.VI from *Jasta* 80b: above at Morsberg in the summer of 1918, below in December 1918 after it was taken over by the USAS. By then the fuselage and nose paintjob had changed somewhat, but the modified lower wing crosses are identical. This may be the aircraft in which Otto Fuchs was photographed. (Peter M. Grosz Collection/STDB)

Above: Pilots from *Jasta* 80b pose in front of what appears to be the *Staffelführer´s* Fokker D.VI 1661/18 for a cheerful group photo. (Peter M. Grosz Collection/STDB)

Above: Around the same time, the groundcrew of *Jasta* 80b gathered in front of a D.VI for their own snapshot. Comparison of the leading edge details indicates that another Fokker served as the backdrop for this picture. (Peter M. Grosz Collection/STDB)

Above: A trio of D.Vis, at least one of them from *Jasta* 80b, seen amongst Fokker D.VIIs and Roland D.VIs after being turned over to the French after the armistice The photo was probably taken at Neunkirchen or Saargemünd. Many surrendered German fighters were gathered here. (Peter M. Grosz Collection/STDB)

Above: This unidentified Fokker D VI was photographed at Lachen-Speyerdorf. The stains on the engine cowling indicate an emergency landing due to an engine defect such as piston seizure. Putting that in context to the location, the aircraft may have been from *Kest* 1a or 1b. (Reinhard Kastner)

Above: This unarmed Fokker D.VI trainer probably served at *Jastaschule* II at Nivelles, where the application of numbers in the style seen here was common. (Peter M. Grosz Collection/STDB)

Above: Fokker D.VI marked with an "X" was operated by an unidentified unit. Interestingly, the fuselage cross remains in its original shape even though a personal marking has been applied to the plane. (Peter M. Grosz Collection/STDB)

Above: This D.VI was apparently evaluated by the French after the war, and it is being pushed back into the hangar. The castor oil stains on the cowling were usually removed by the groundcrew after each flight. But since the evaluation was probably fairly short the new owners did not bother with this. (Peter M. Grosz Collection/STDB)

Above: This Fokker D.VI in French hands after the war may be the same aircraft pictured above. At this point, the machine guns had been removed. On this D.VI, the rudder cross has been slimmed down to the final 8:1 proportions, which was not always done on older aircraft. This may be a hint that this particular D.VI actually saw use until the final days of the war, which was the case with a few Fokker Triplanes as well. On fall days with a low cloud ceiling, the rotary-engined Fokker fighters were still rather useful. (Greg VanWyngarden)

Right: A pilot of *Jasta* 80b posing with a Fokker D.VI of *Jasta* 80b. This is the second aircraft seen in the earlier *Jasta* 80b lineup photo. (Museum der Pfalz via Greg VanWyngarden)

Late Fokker Fighters in Retrospect

Recorded Time to Climb in Minutes								
Climb to:	Pfalz Dr.I	Dr.I	V17	V25	D.VI	D.VII	D.VIII	V29
1000 m	1.5	2.9	3.2	1.5	2.6	1.9	2.0	—
2000 m	3.0	5.5	6	3.3	5.9	4.0	4.5	—
3000 m	5.5	9.3	9	6.5	10.7	7.0	7.5	7.5
4000 m	8.5	13.9	13	9.5	17.3	10.2	10.8	10.5
5000 m	11.5	21.9	19	13.3	28.0	14.0	15.0	14.7
6000 m	14.5	—	—	—	—	18.7	19.5	19.7

1. The Pfalz Dr.I triplane recorded the fastest climb rate of any WWI German aircraft. It was powered by a 160 hp Sh.III counter-rotary engine. On test the engine actually developed about 205 hp.
2. The Fokker Dr.I, V17, V25, D.VI, and D.VIII all used the 110 hp Oberursel Ur.II. The V25 was also tested with the 170 hp. Goebel Goe.III but the data are for the Ur.II. The Ur.III was installed in the production batch of D.VIII fighters delivered starting 8 October 1918, but the data are for the Ur.II.
3. The times for the D.VII are for the 185 hp over-compressed BMW IIIa . The V29 used the same engine.
4. The V17 was the fastest fighter at the First Fighter Competition but its exact speed is not known. The similar V25 was clocked at 200 km/hr., as was the D.VII, D.VIII, and V29.
5. Performance measurements are not exact; they depend on air temperature and pressure, the ability of the pilot, condition of the airframe and engine, and measurement error. Time to climb is recorded by watch and maximum airspeed is more difficult to determine.
6. Engine power depends on air density, which drops with altitude so engine power drops with altitude.
7. Climb rate depends directly on the amount of excess power available (the power in excess for that needed for level flight).
8. Maximum speed for a given airframe depends on the cube root of the power available. In other words, to double the airspeed requires 8 times the power for a given airframe.

The Fokker Dr.I Triplane was the first production Fokker fighter after the failure of the early Fokker biplane fighters. All the Fokker fighters listed here, prototype or production, exhibited some common design features. Most important was Fokker's advanced thick-airfoil wooden wing.

Other common features were welded steel tube construction of the fuselage and tail surfaces covered with fabric, an undercarriage airfoil over the axle, and mounting of two synchronized machine guns.

All production Fokkers were maneuverable and had good flying qualities, especially the later types with Fokker's thick wooden wings. Of course, the smaller, lighter, rotary-powered types tended to be a little more agile. Combat effectiveness depended also on speed, climb rate, and the pilot's field of view.

The above table of recorded climb times is provided for reference. All the fighters are late Fokker types except for the Pfalz Dr.I, which is provided as the standard because it recorded the fastest climb of any German aircraft during WWI. Its powerful Sh.III counter-rotary took some time to develop but when running well gave all the fighter types it powered an excellent climb rate.

First to the front in the autumn of 1917 was the Fokker Dr.I Triplane. Relatively slow due to its triplane configuration (the actual speed has given in the range of 160–185 km/hr, but the lower range is more likely), the Dr.I had an excellent rate of climb at low altitude but it was only about half that of the more powerful but less maneuverable Pfalz Dr.I. The Fokker Dr.I depended on its extreme maneuverability and excellent low-altitude climb rate, but was limited by its low speed.

The Fokker V17 and the prototypes of the D.VI and D.VII competed at the First Fighter Competition at the beginning of 1918. The V17 monoplane was the fastest fighter in the competition and demonstrated excellent climb. Nevertheless, it was not ordered into production; instead the D.VI and D.VII were placed in production despite their inferior performance. The reason given for not producing the V17 was that the wing obstructed the pilot's field of view downward. There was some truth to this, but the real reason was probably anxiety over its monoplane configuration. After all, the Fokker E-type fighters all had this configuration.

The V25 monoplane was tested after the V17. The main difference between the types was the position of the wing. The V25 had its wing mounted at the bottom of the fuselage and thus offered an improved view for the pilot. The V25 was tested with both the 110 hp Oberursel Ur.II and the more powerful 170 hp Goebel Goe.III engine. Even the

110 hp engine gave it excellent performance as shown in the table, but again the pilot's field of view was not considered satisfactory and it was not developed further. Again, it must be emphasized that the widely-used Fokker E-type fighters shared this configuration.

The low position of the monoplane wing was almost standard for WW2 fighters and the Junkers D.I with the same low monoplane wing position was placed in production, so one suspects the pilot's field of view issue was not as serious as *Idflieg* stated. Again, the V17 and V25 monoplanes possessed excellent speed, climb, and maneuverability and *Idflieg's* decision to abandon them without even ordering a modest production batch for combat evaluation seems to be one of *Idflieg's* more serious mistakes.

The V25 had a much better rate of climb than the very similar V17 that is not explainable by the information available. Both used the same engine and were similar in size, weight, and configuration. Perhaps the difference was due to engine output, pilot skill, or different conditions, but could be due to refinements to the V25 airframe not apparent in the photos, such as better airfoil, propeller, etc.

The V9 and V13 lead to the D.VI production fighter that was fast and had a good rate of climb at low altitude. But the D.VI rate of climb fell behind the Dr.I above 2,000 m, largely explaining its small production order despite its higher speed. Since the Dr.I and D.VI both used the Ur.II, what explains the major difference in high-altitude climb rate? Again, engine output, pilot skill, and different conditions suggest themselves as factors, but they do not explain the significant difference because one assumes the trials were made with relatively good engines and skilled pilots. The D.VI did have wings of lower aspect ratio that would generate more drag at the higher angles of attack that would be needed as the aircraft climbs through the thinner air at higher altitude. Whatever the reason for the poor climb at altitude, the D.VI did not meet the combat needs of 1918.

The larger D.VII powered by a 170 hp Mercedes D.IIIa inline engine, later a high-altitude 185 hp BMW IIIa , offered a much better climb rate than the rotary-powered D.VI. It also had wings of higher aspect ratio that helped with the climb rate. Combined with the BMW engine the Fokker D.VII offered excellent speed, climb, and maneuverability at all altitudes.

Eliminating the problem of poor field of view, the D.VIII parasol monoplane finally gained production approval and offered excellent speed and climb despite its low-power Ur.II. The lower drag and weight of the D.VIII made it much faster and gave a much better climb rate than the Triplane or D.VI biplane that used the same engine type.

The final step in WWI Fokker fighter development was the BMW-powered V29, a parasol monoplane development of the Fokker D.VII biplane. Probably a little faster than the D.VII (actual specifications are not known), the V29 had a slightly lower climb rate likely due to its smaller wing area and higher wing loading. Although the V29 won the Third Figher Competition, it offered little advantage over the BMW-powered D.VII biplane. Performance of the two types was very similar, the D.VII was already serving with great distinction at the front, and its biplane wing cellule was proven and offered a stiffer wing for better aileron response. With the recent wing failures of the D.VIII parasol monoplane fresh in mind and the terrible war situation a few weeks before the Armistice, no orders for the V29 were placed. Finally, the large number of surplus D.VII fighters available after the war eliminated any desire to produce the V29.

Other late Fokker prototypes were variations of existing production types that offered no significant improvement. Even the C.I two-seater development of the D.VII was much more difficult to develop into a satisfactory production aircraft than expected.

Fokker's later fighters depended on his thick wooden wing technology for their success. This was the critical breakthrough development that was key to the performance and flying qualities of his late-war fighters. No WWI Allied fighter, which all used thin airfoils, could match the excellent flying qualities of the late-war Fokkers.

Selected Bibliography

Books

Bodenschatz, Karl, Jagd In Flanderns Himmel, Verlag Knott & Hirth, München, 1935

Braatz, Kurt, Robert von Greim Band I, NeunundzwanzigSechs Verlag, Moosburg, 2016

Bronnenkant, Lance, PhD., *The Blue Max Airmen*, Vols. 1–19, Aeronaut Books, 2012–2022.

Dierikx, Marc I. J., *Fokker | A Transatlantic Biography*, Smithsonian Institution, Washington, DC, 1997.

Engels, Achim Sven, *Fokker Und Seine Flugzeuge*, Fokker Team Schorndorf, 1996.

Fokker, Anthony H.G., *Flying Dutchman*, Penguin Books, London, 1938.

Gray, Peter, and Thetford, Owen, *German Aircraft of the First World War*, second revised edition, New York: Doubleday & Company, Inc., 1970.

Green, William, and Swanborough, Gordon, *The Complete Book of Fighters*, Smithmark Publishers, 1994.

Grosz, P.M., *Fokker D.VI*, Albatros Productions, Berkhamsted, 2000.

Grosz, P.M., *Fokker D.VII*, Albatros Productions, Berkhamsted, 1989.

Grosz, P.M., *Fokker D.VIII*, Albatros Productions, Berkhamsted, 1991.

Grosz, P.M., and Volker Koos, *Fokker Flugzeugwerke in Deutschland 1912–1921*, HEEL Verlag, GmbH, 2004.

Herris, Jack, *The German Fighter Competitions of 1918*, Aeronaut Books, 2013.

Imrie, Alex, German Fighter Units June 1917-18, Osprey Publishing, 1978

Imrie, Alex, German Air Aces Of World War One, Arms & Armour Press, 1987

Lange, Bruno, *Das Buch der Deutschen Luftfahrtechnik*, Verlag Dieter Hoffman, Mainz, 1970.

Möller, Hanns, Kampf und Sieg, Verlag Bernard & Graefe, Berlin, 1939.

Postma, Thijs, *Fokker | Aircraft Builders to the World*, Jane's Incorporated, New York, 1980.

Rimell, Ray, ed., *Fokker D.VII Anthology 1*, Albatros Productions, Berkhamsted, 1997.

Rimell, Ray, ed., *Fokker D.VII Anthology 2*, Albatros Productions, Berkhamsted, 2000.

Rimell, Ray, ed., *Fokker D.VII Anthology 3*, Albatros Productions, Berkhamsted, 2001.

Schmäling, Bruno, and Bock, Winfried; *Royal Prussian Jagdstaffel 30*, Aeronaut Books, Indio, 2014.

Schmäling, Bruno, and Bock, Winfried; *Royal Bavarian Jagdstaffel 23*, Aeronaut Books, Reno, 2018.

Schmäling, Bruno, and Leckscheid, Jörn; *Jasta Colors Volume 1*, Aeronaut Books, Reno, 2020.

Schmäling, Bruno, Kastner, Reinhard, and Leckscheid, Jörn; *Royal Bavarian Jagdstaffel 76*, Aeronaut Books, Reno, 2022.

Stark, Rudolf, Die Jagdstaffel unsere Heimat, v. Hafe & Koehler Verlag Leipzig, 1932

VanWyngarden, Greg, von Richthofen´s Flying Circus, Albatros Productions, Berkhamsted, 1994.

Weyl, A.R., *Fokker | The Creative Years*, Putnam, London, 1965.

Zuerl, Walter, Pour le mérite-Flieger, Curt Pechstein Verlag, München, 1938

Articles

Grosz, Peter M., "*Frontbestand*" *WW1 Aero* No.107, Dec. 1985, p.60–66.

Grosz, Peter M., "*Frontbestand*" *WW1 Aero* No.108, Feb. 1986, p.66–69.

Phillips, Terry H., "The Fokker V.7 Experimental Triplanes", *Over the Front* No.22-3, Fall 2007, p.196–217.

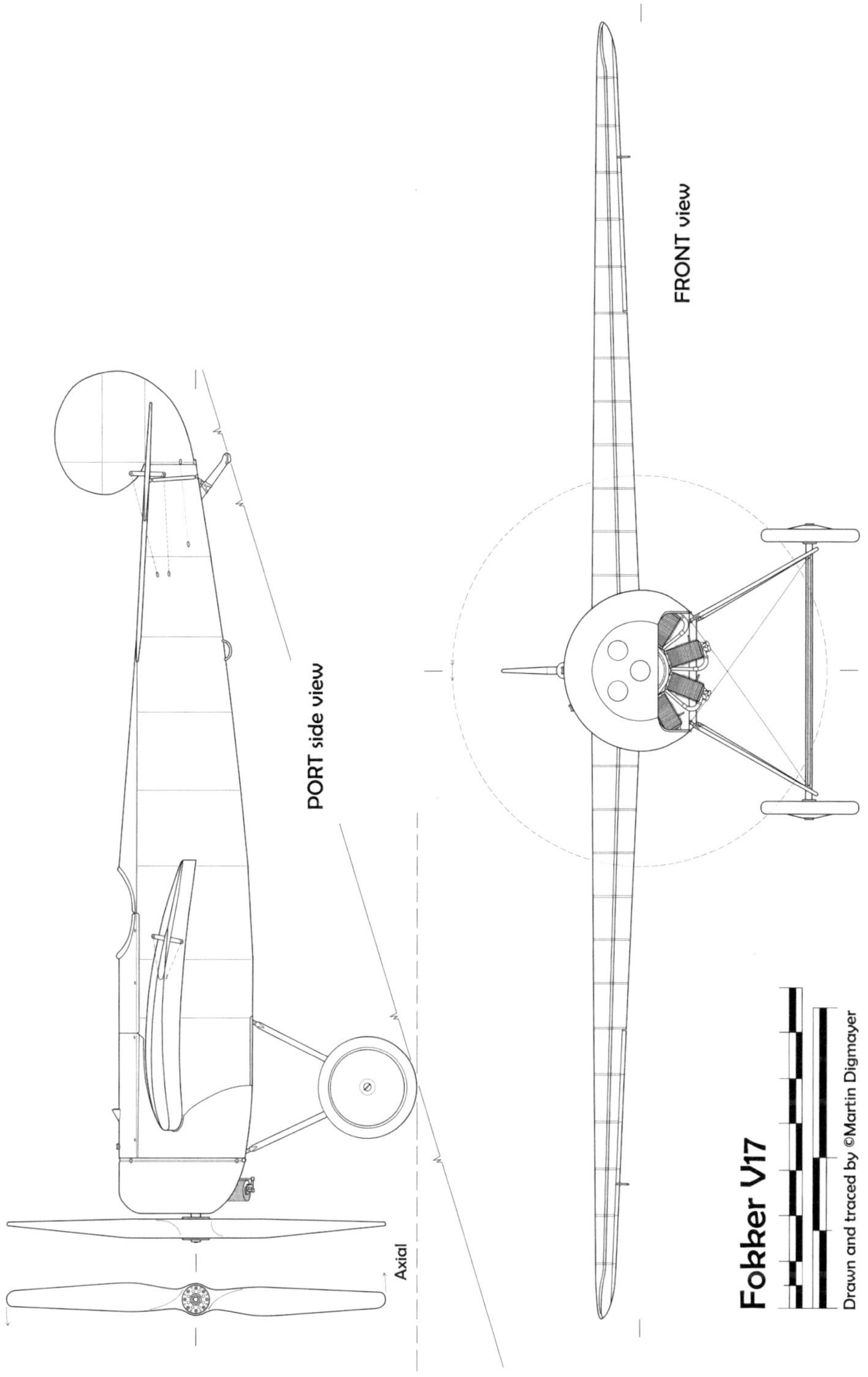
PORT side view
FRONT view
Axial
Fokker V17
Drawn and traced by ©Martin Digmayer

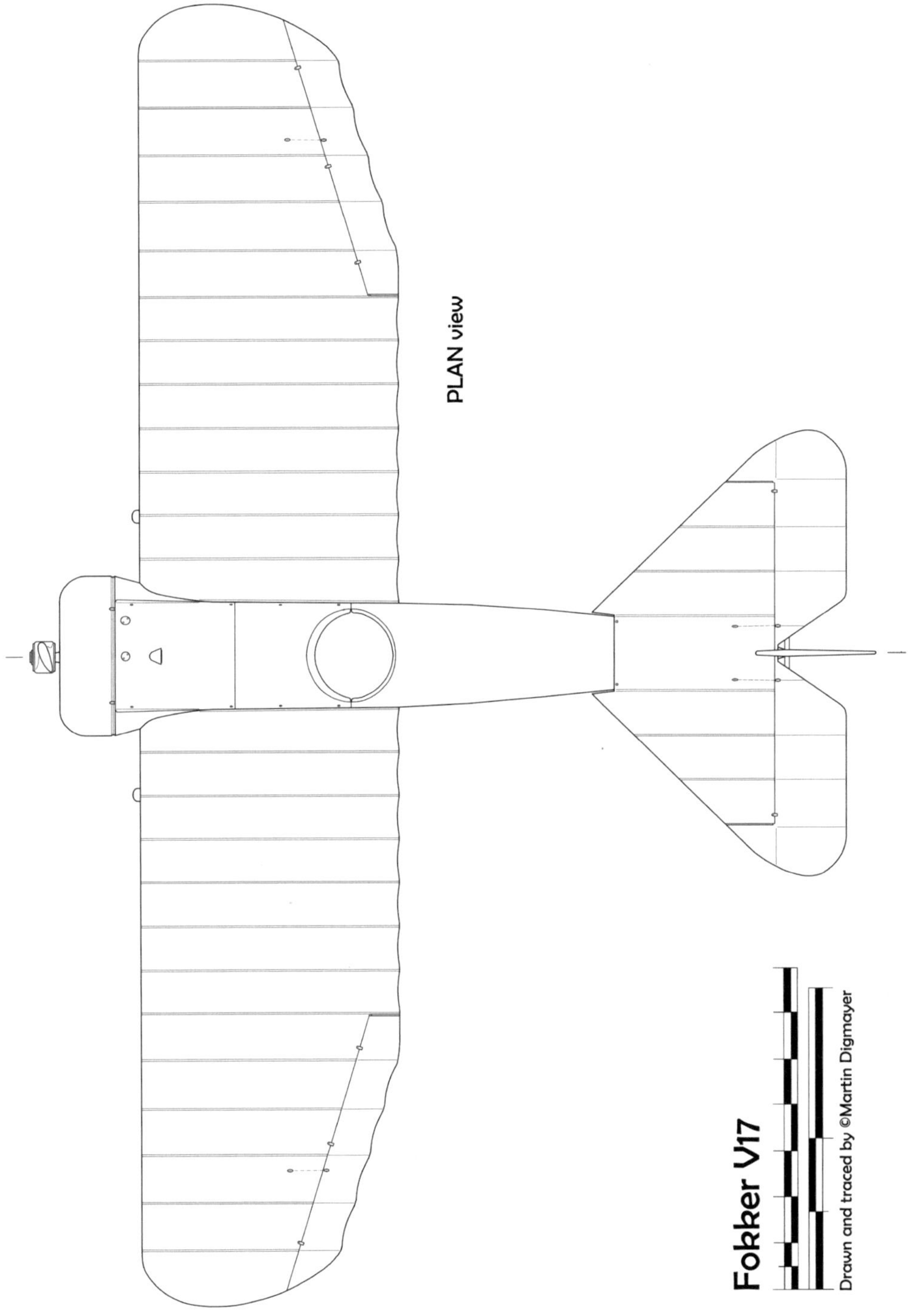
PLAN view
Fokker V17
Drawn and traced by ©Martin Digmayer

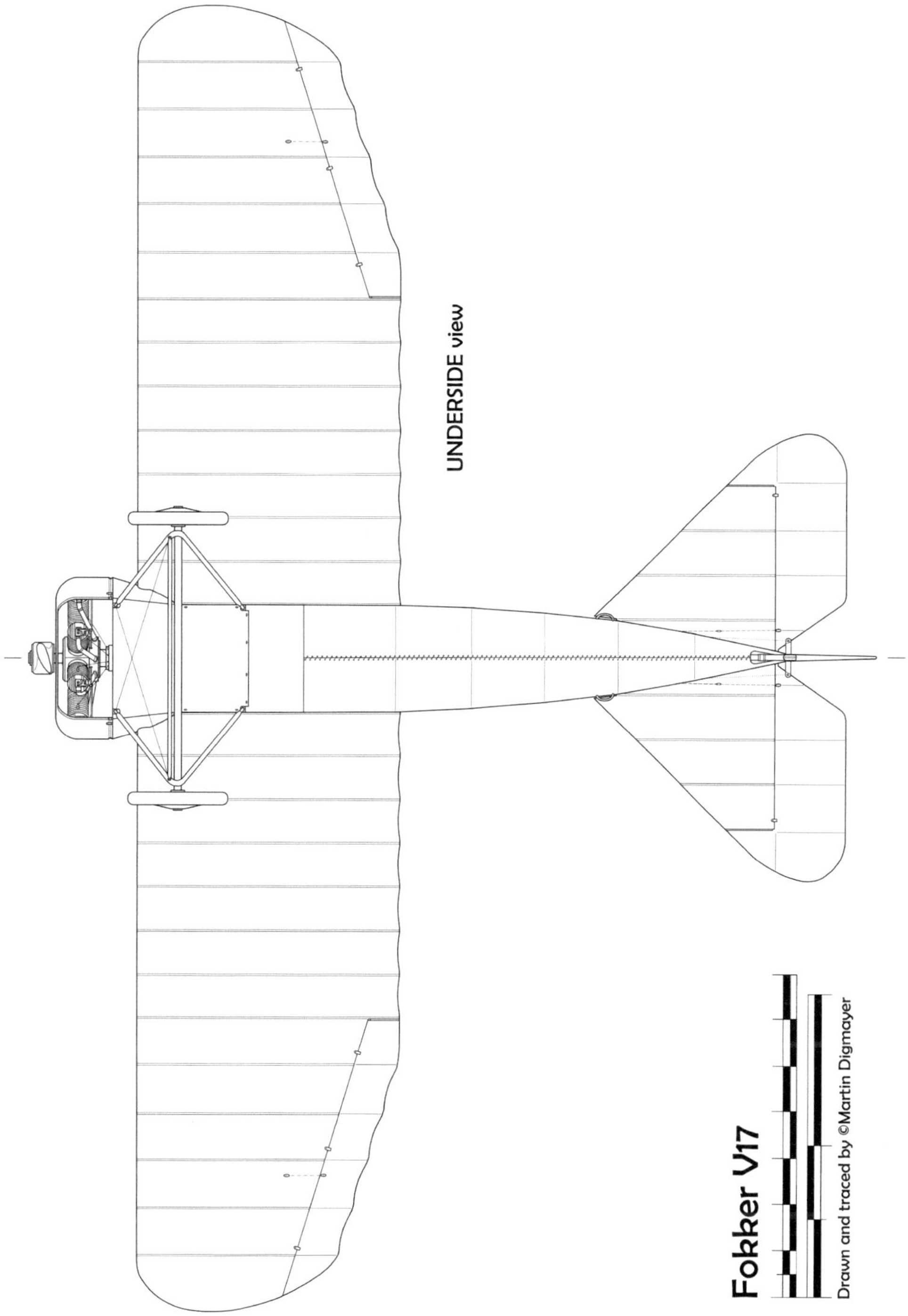
UNDERSIDE view
Fokker V17
Drawn and traced by ©Martin Digmayer

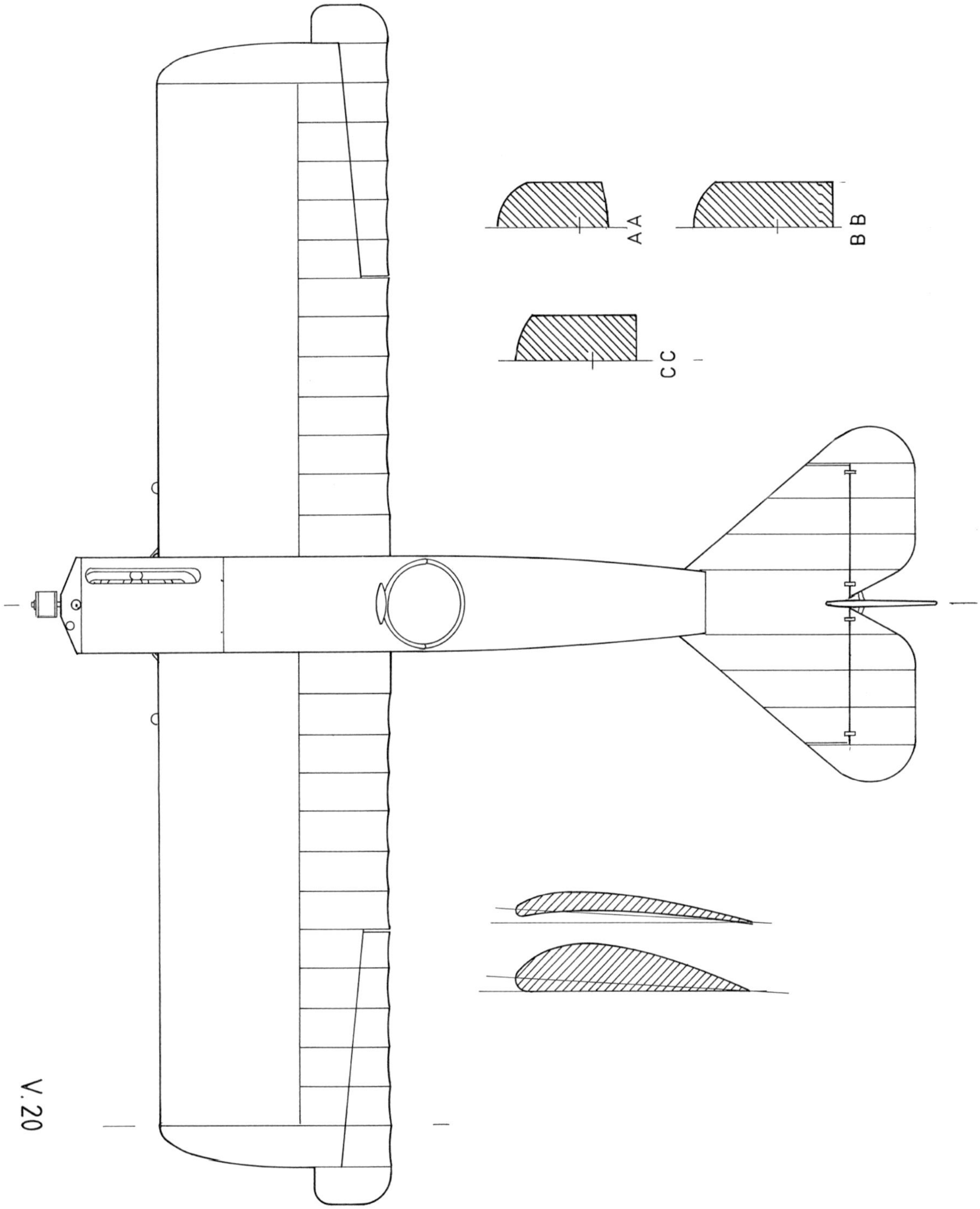
A A
B B
C C
V.20

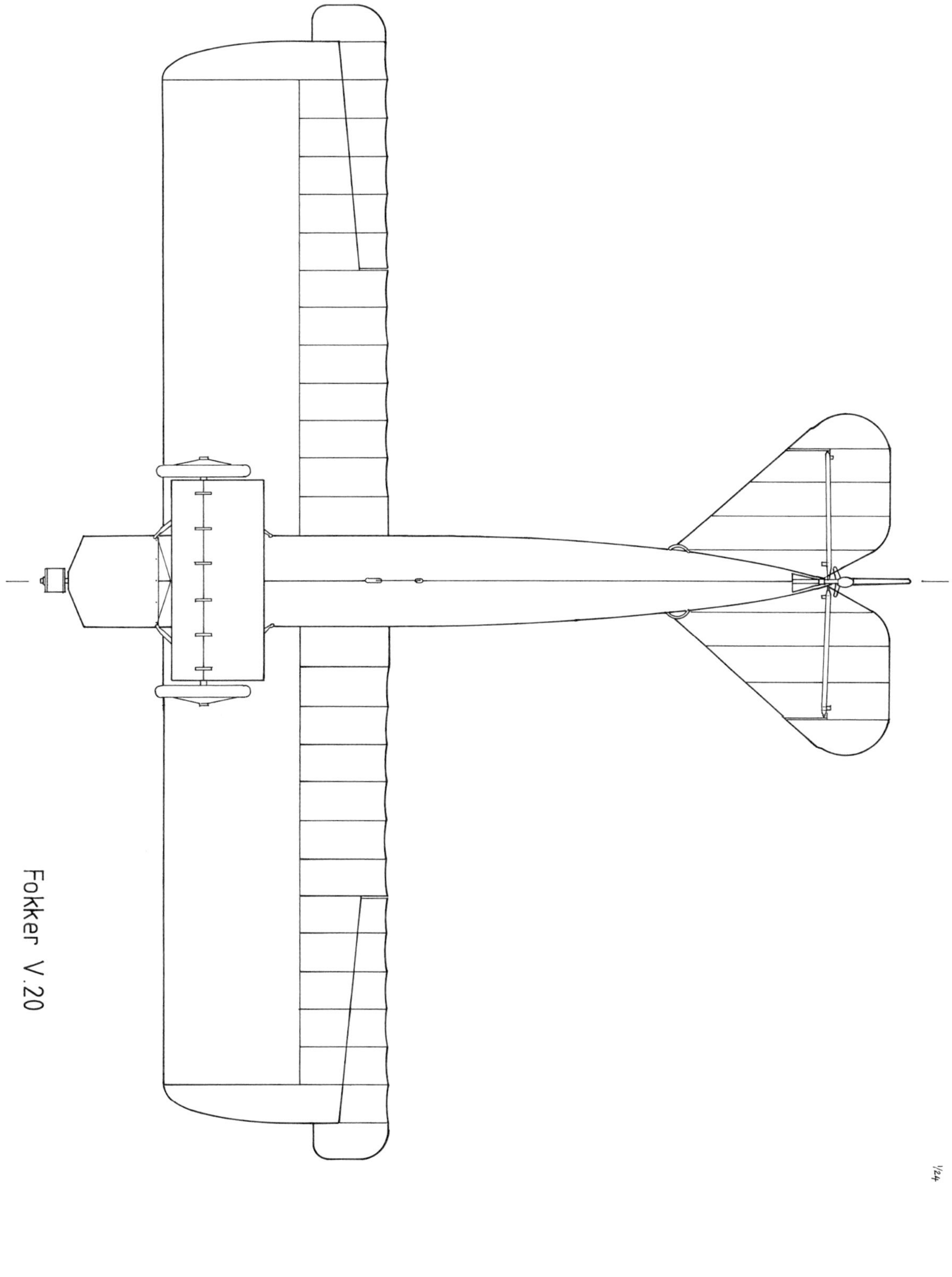
Fokker V.20
1/24

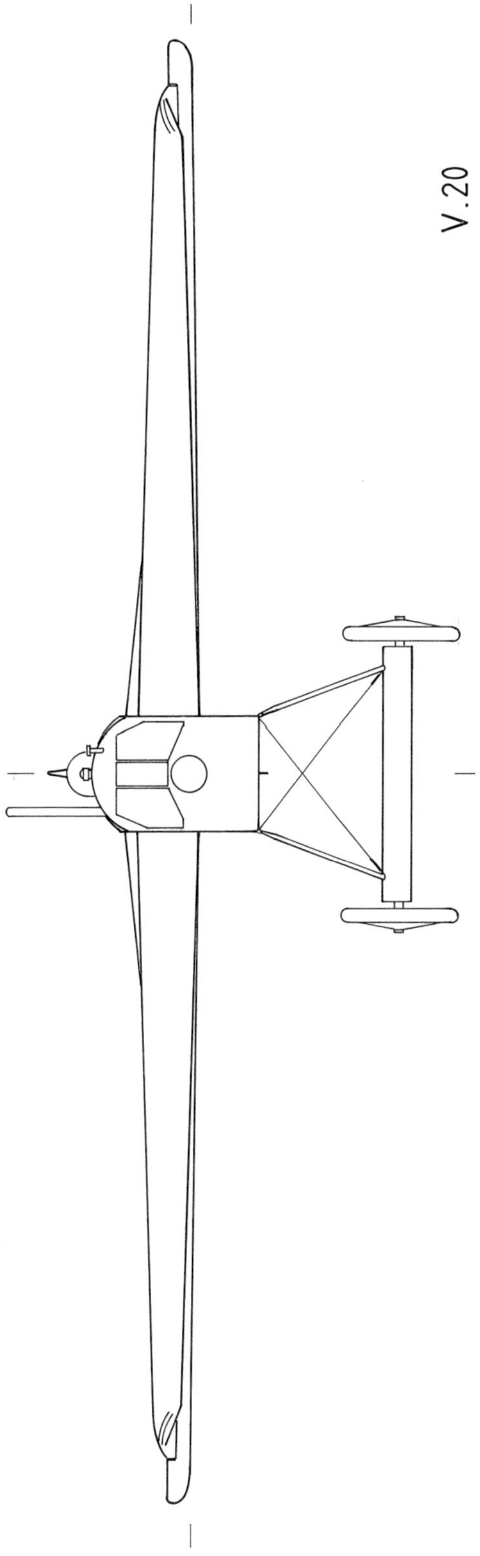
V.20

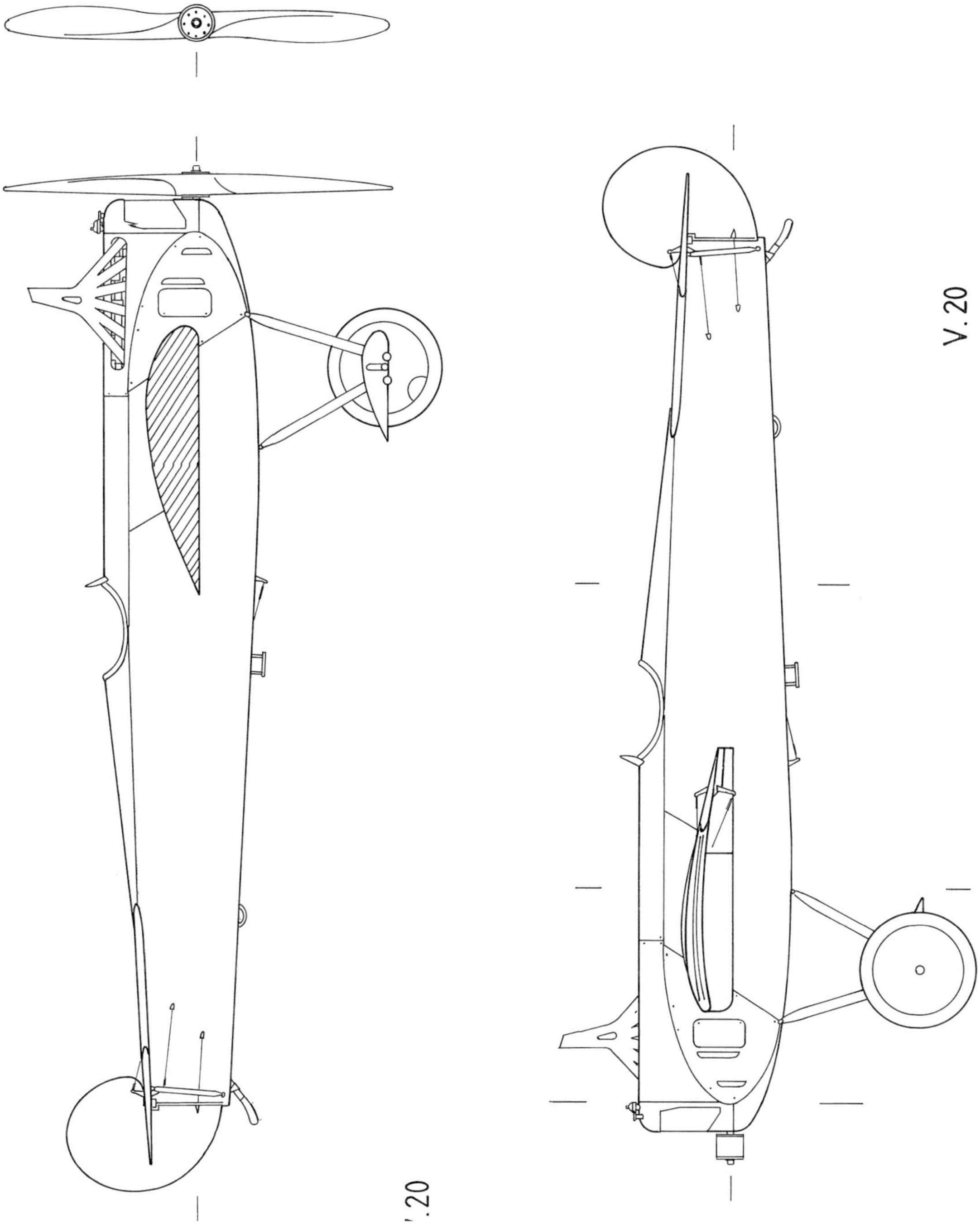
V. 20

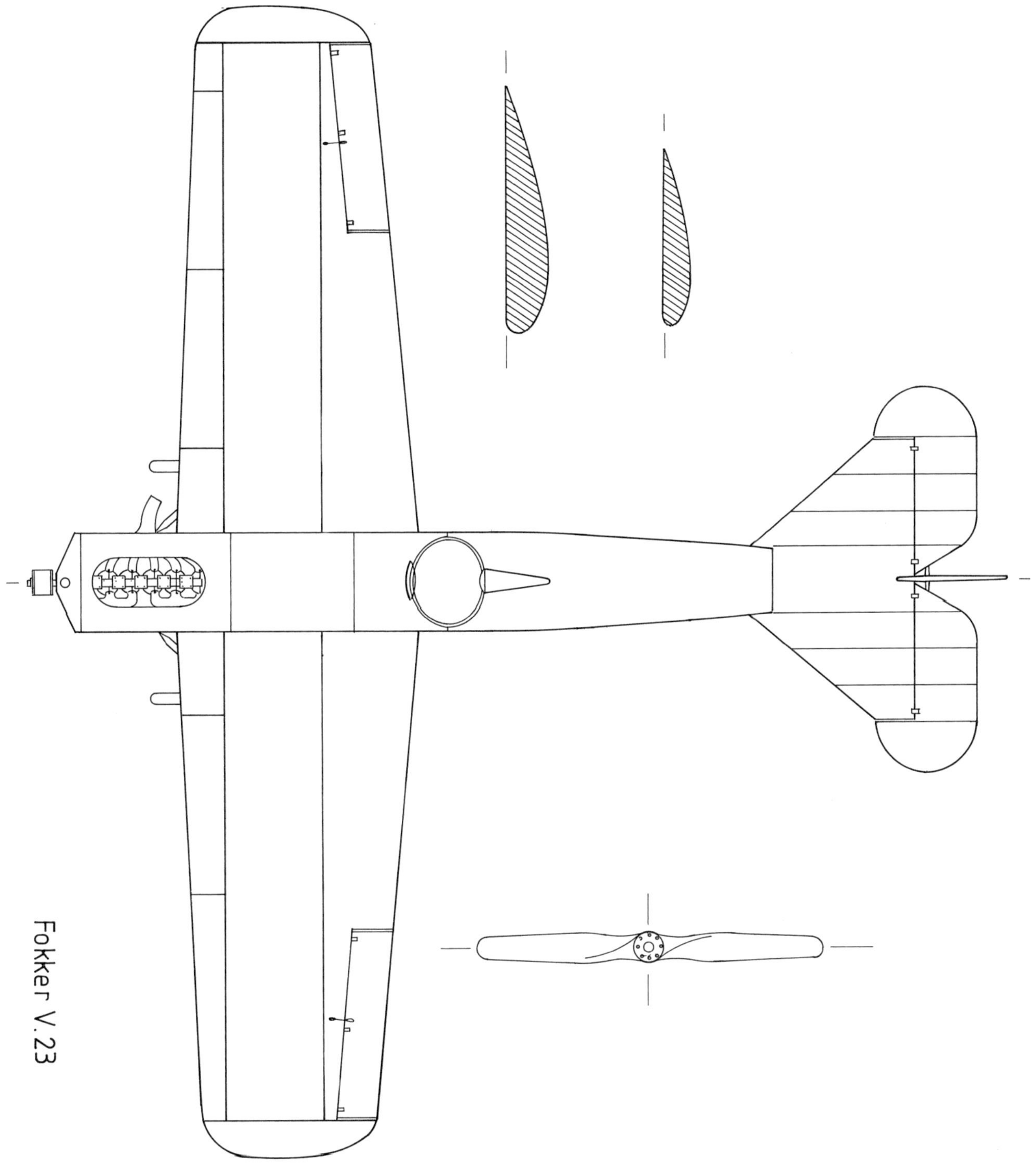

Fokker V.23

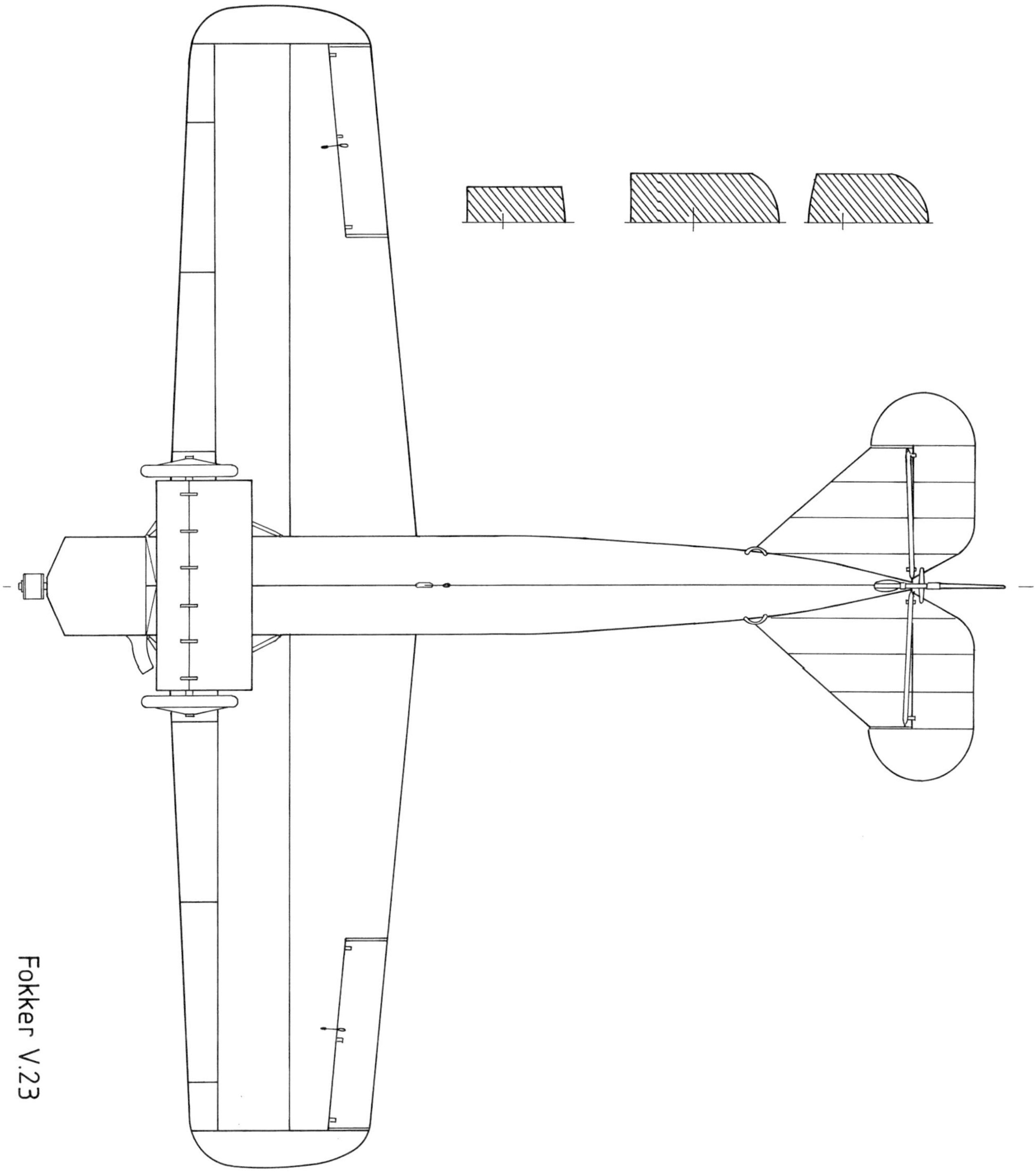

Fokker V.23

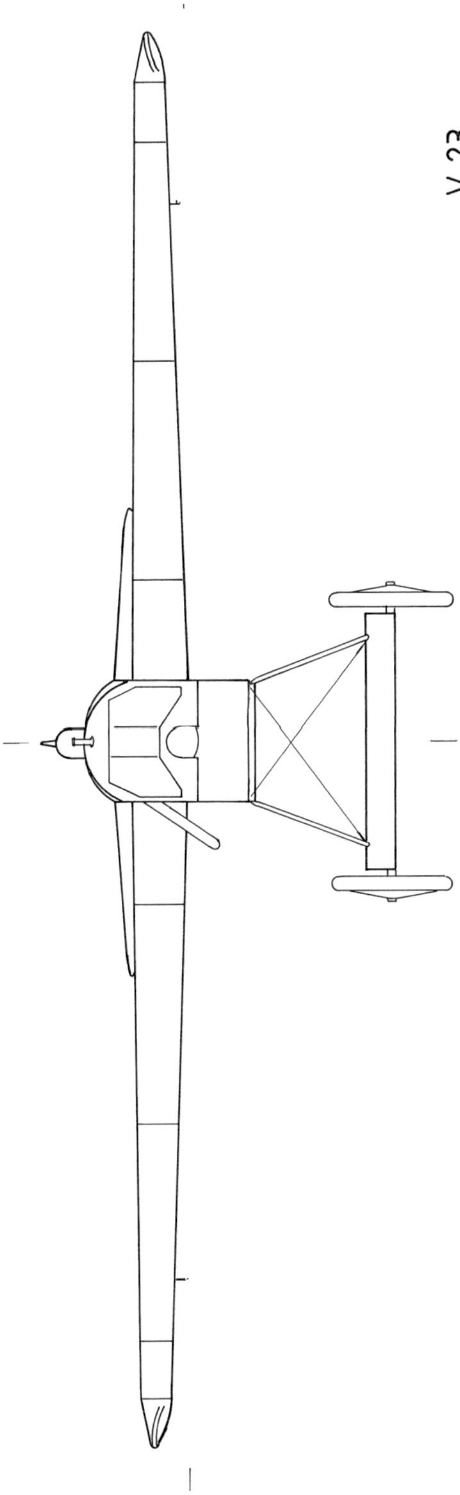

V.23

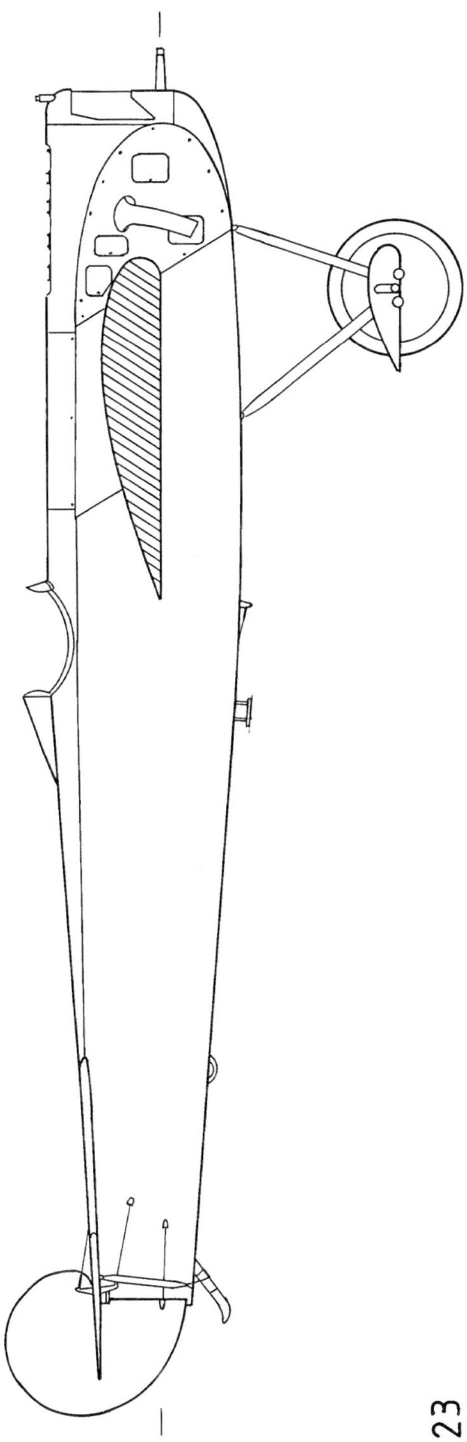

V.23

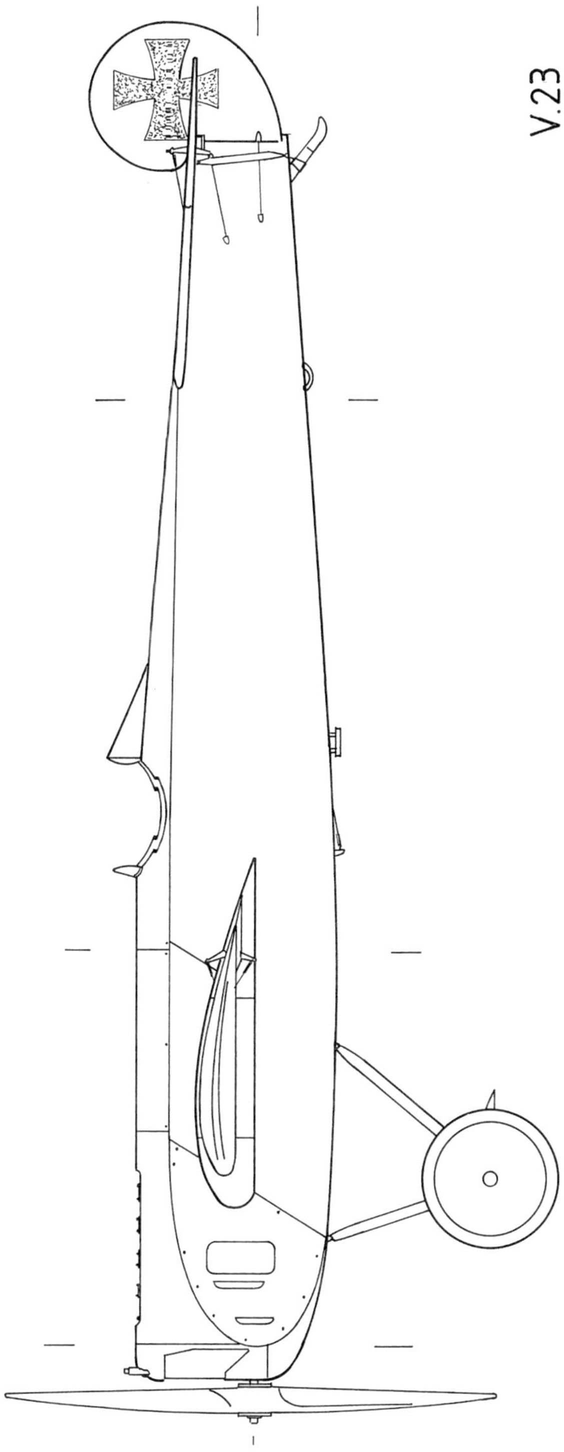

V.23

Fokker V.25

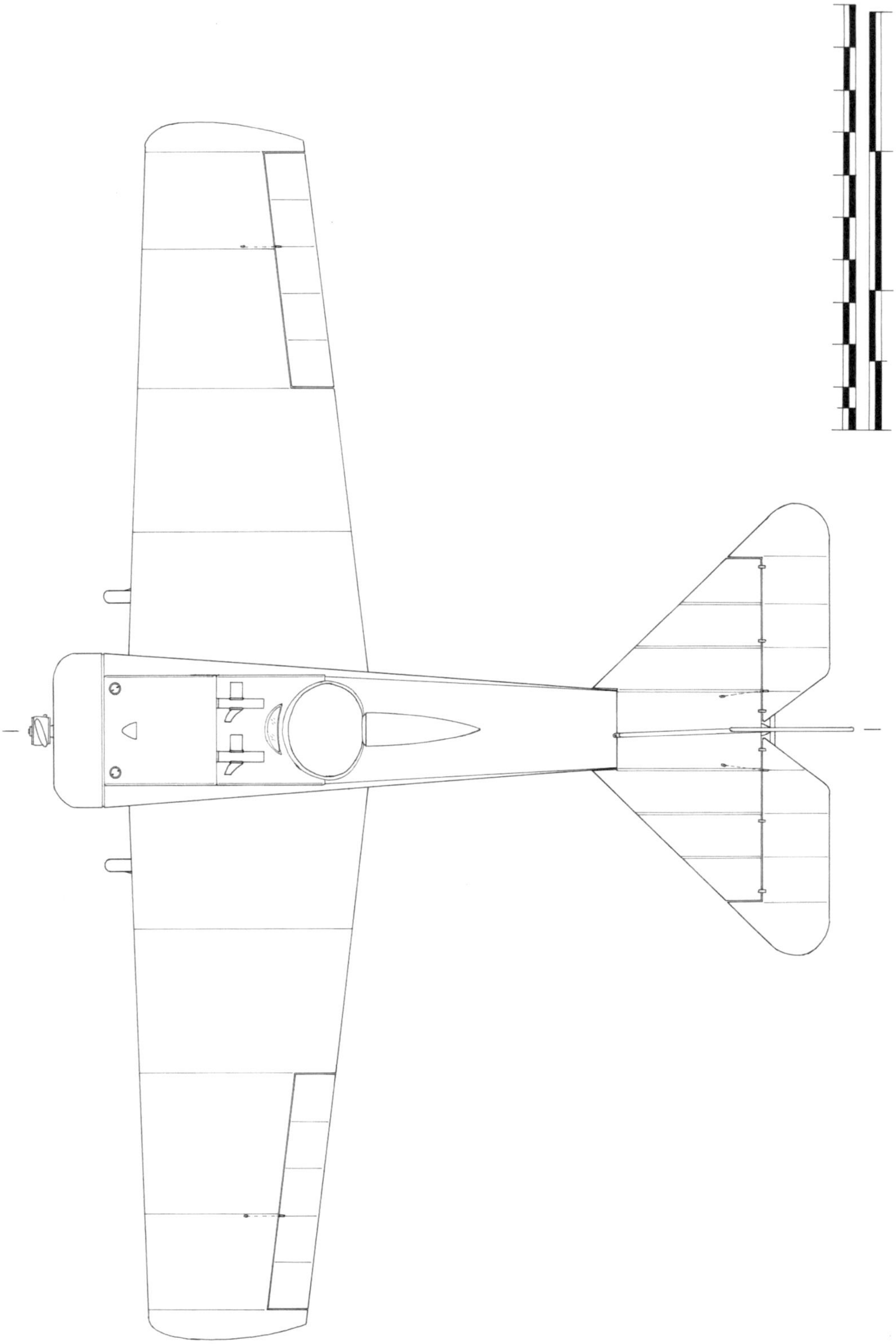

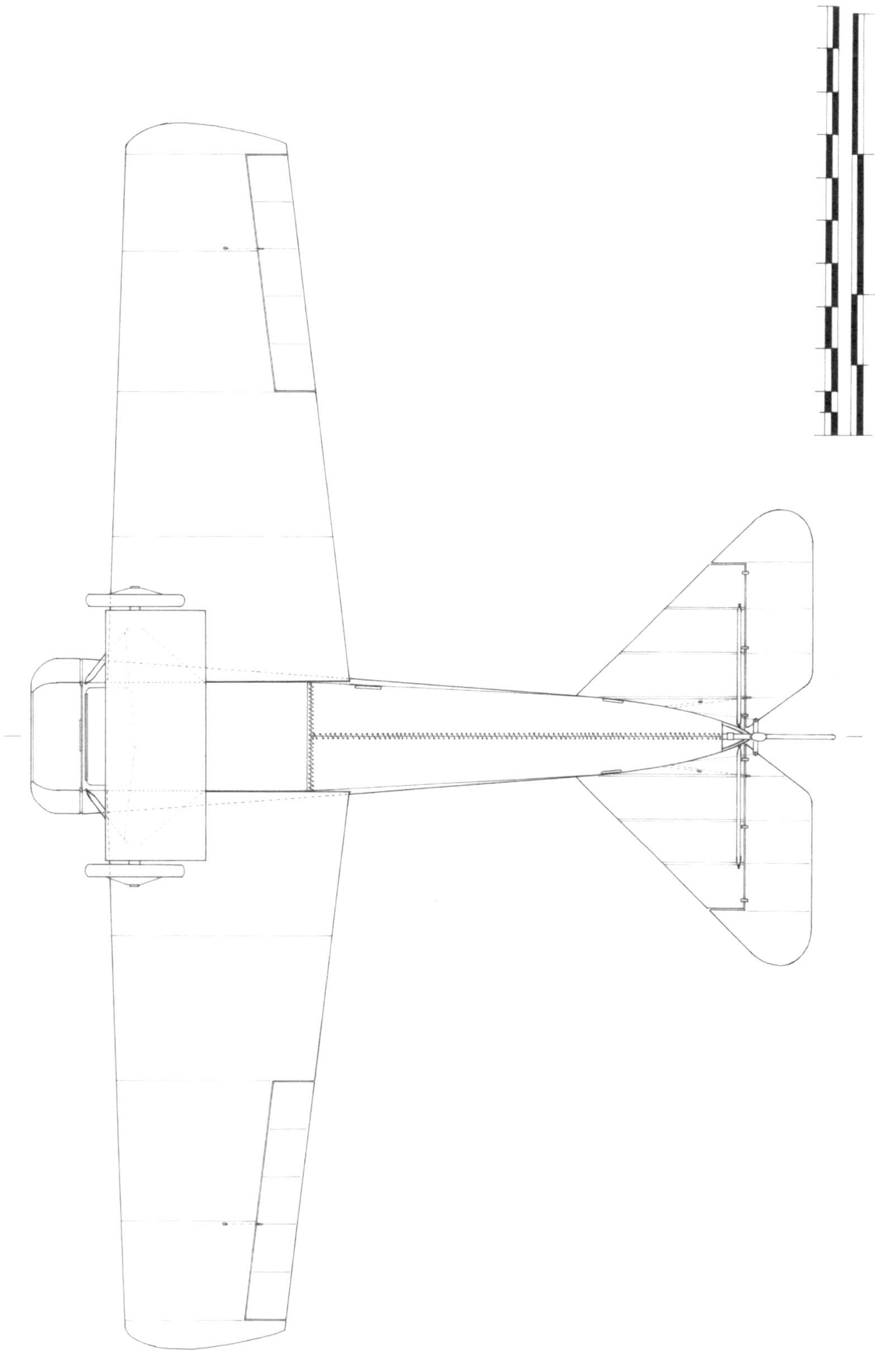

Fokker V.25

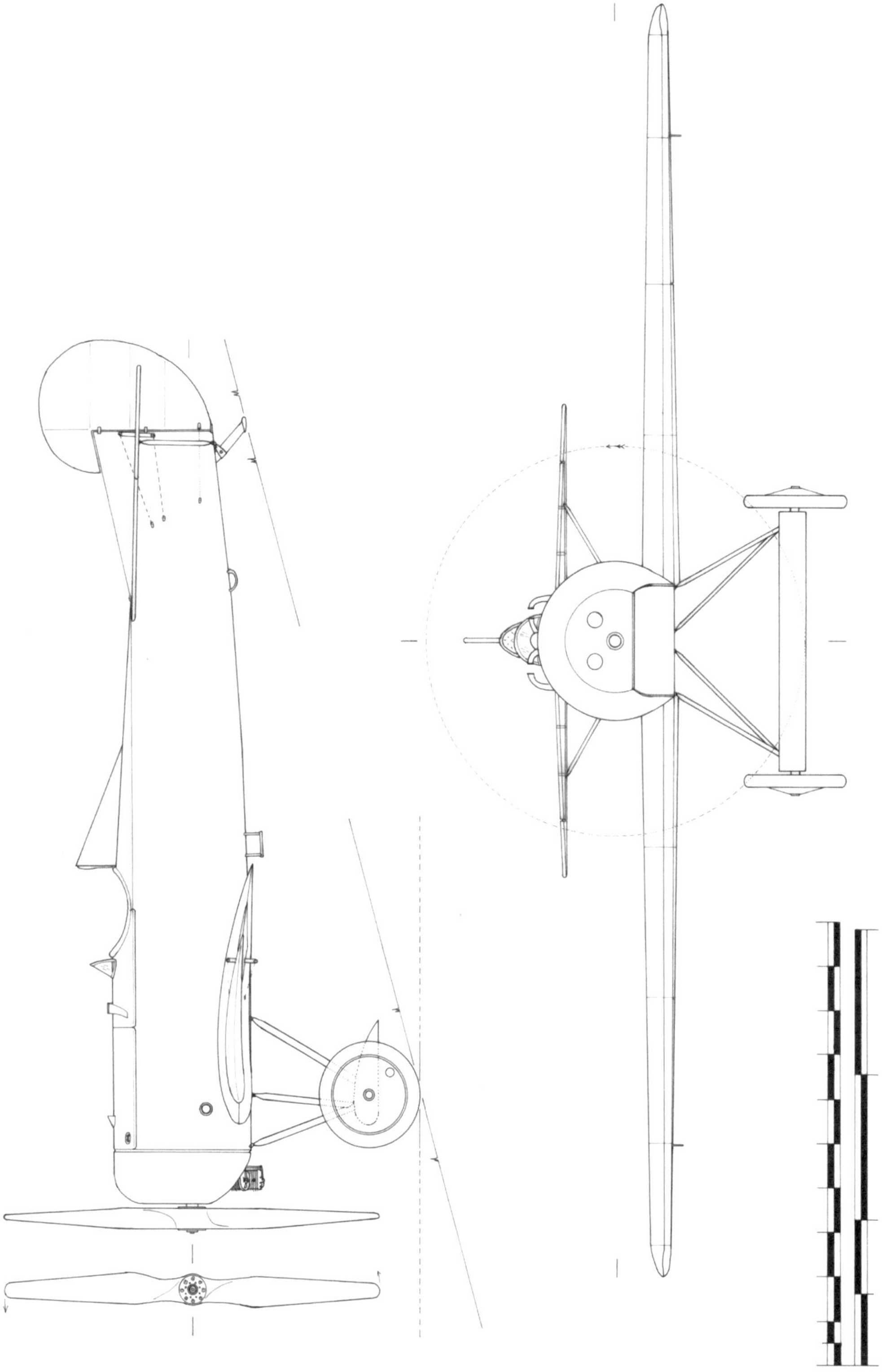

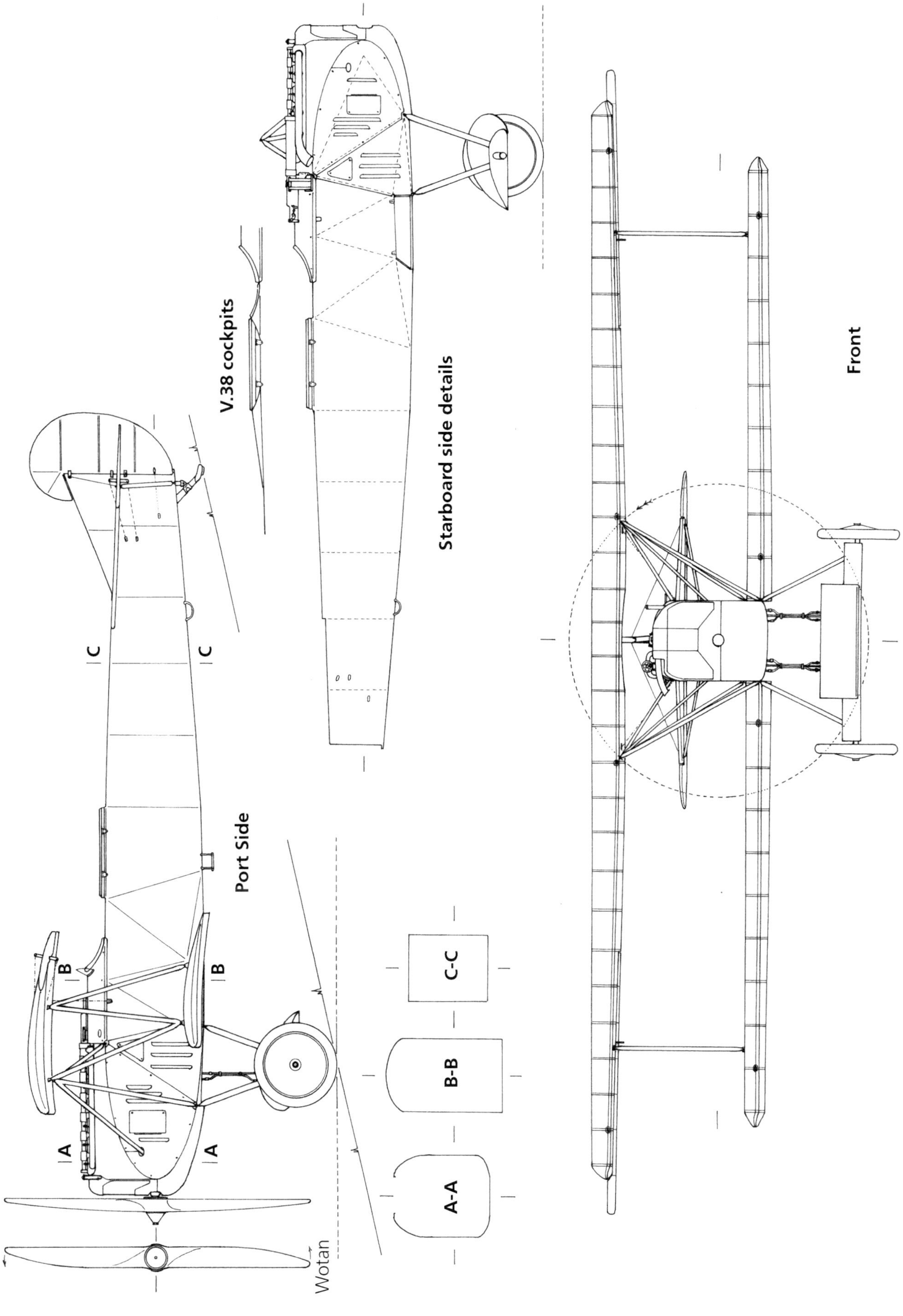
V.38 cockpits
Starboard side details
Front
Port Side
C
C
B
B
A
A
C-C
B-B
A-A
Wotan

Fokker V.38 / C.I

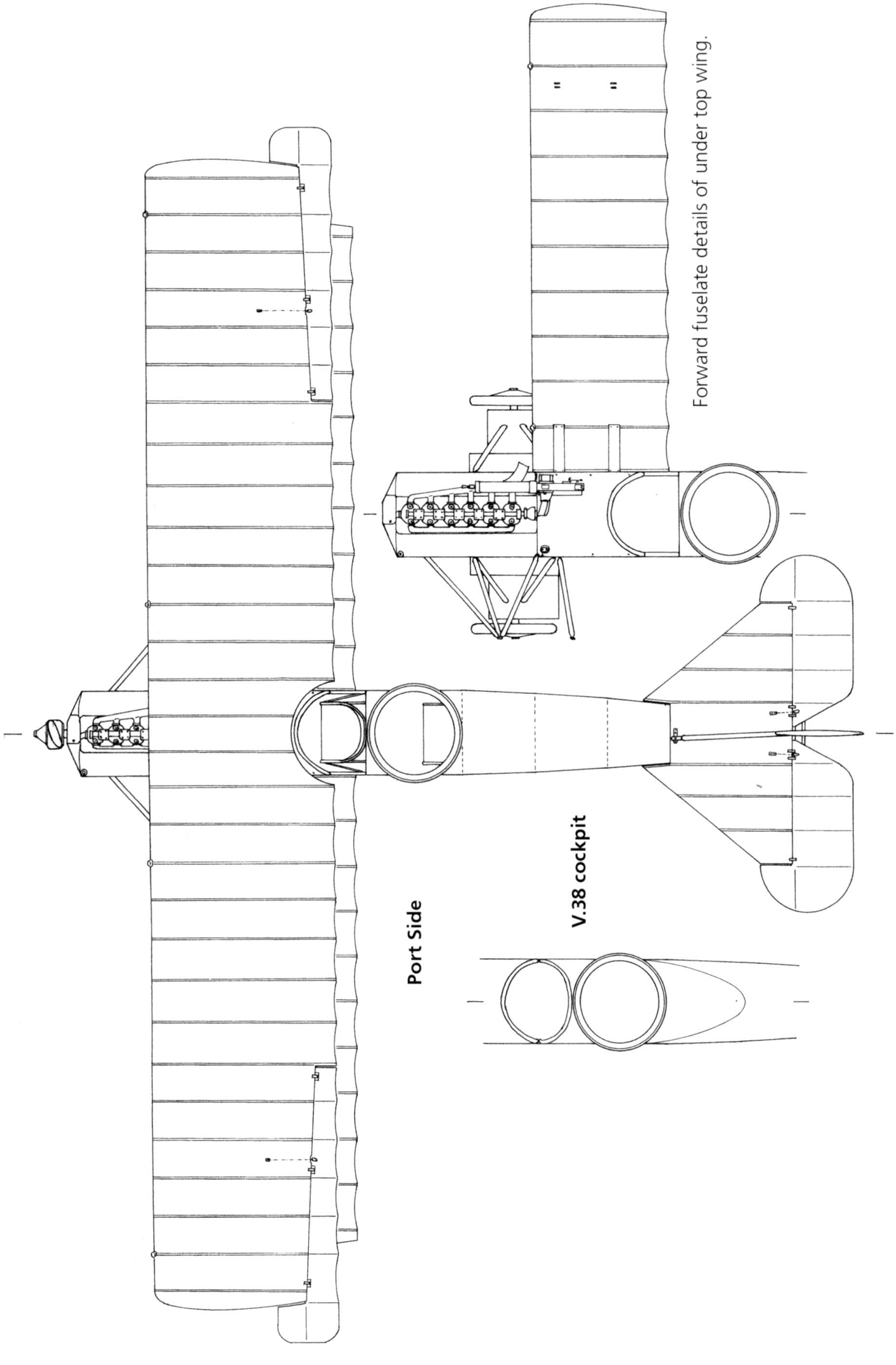

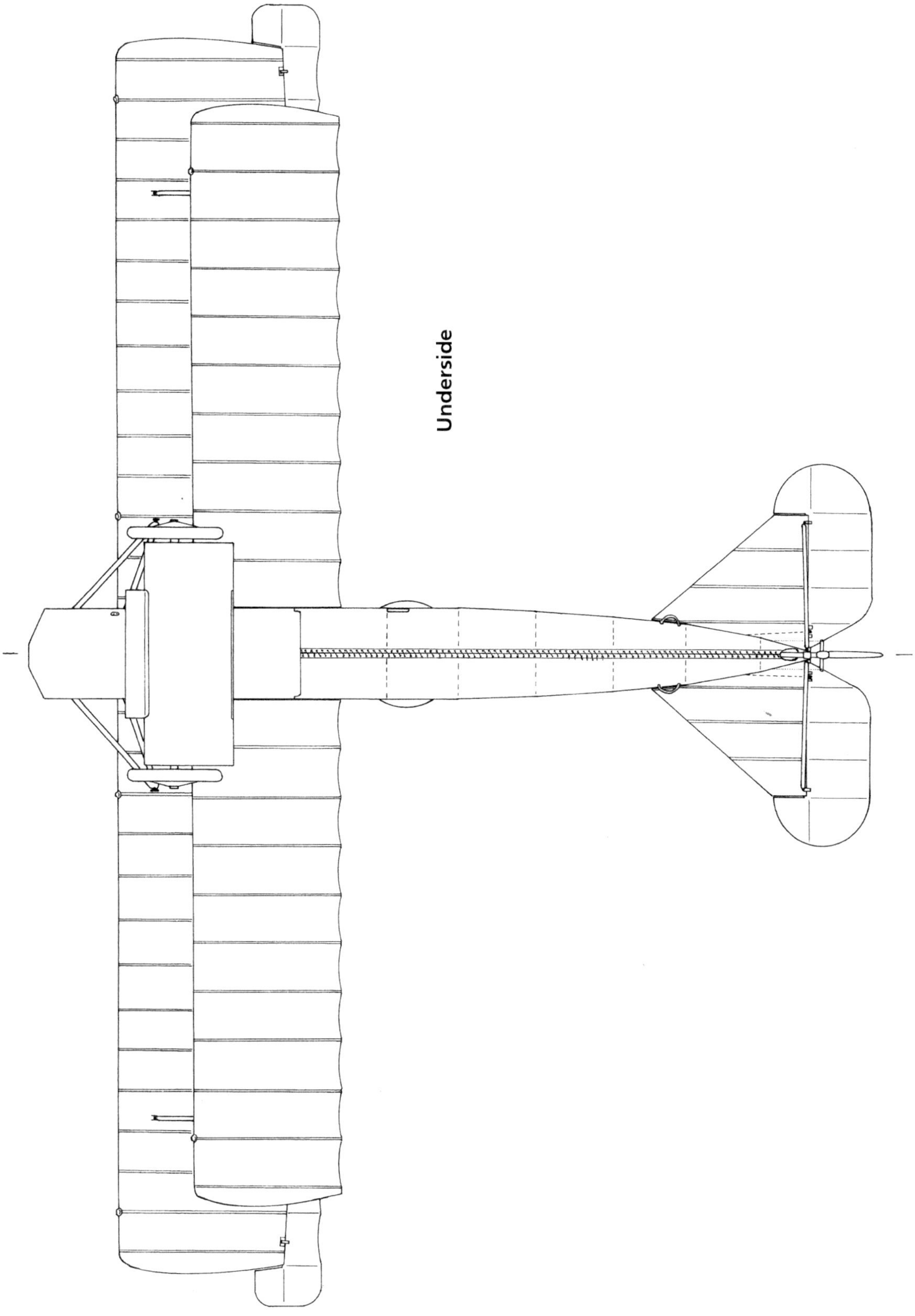
Underside

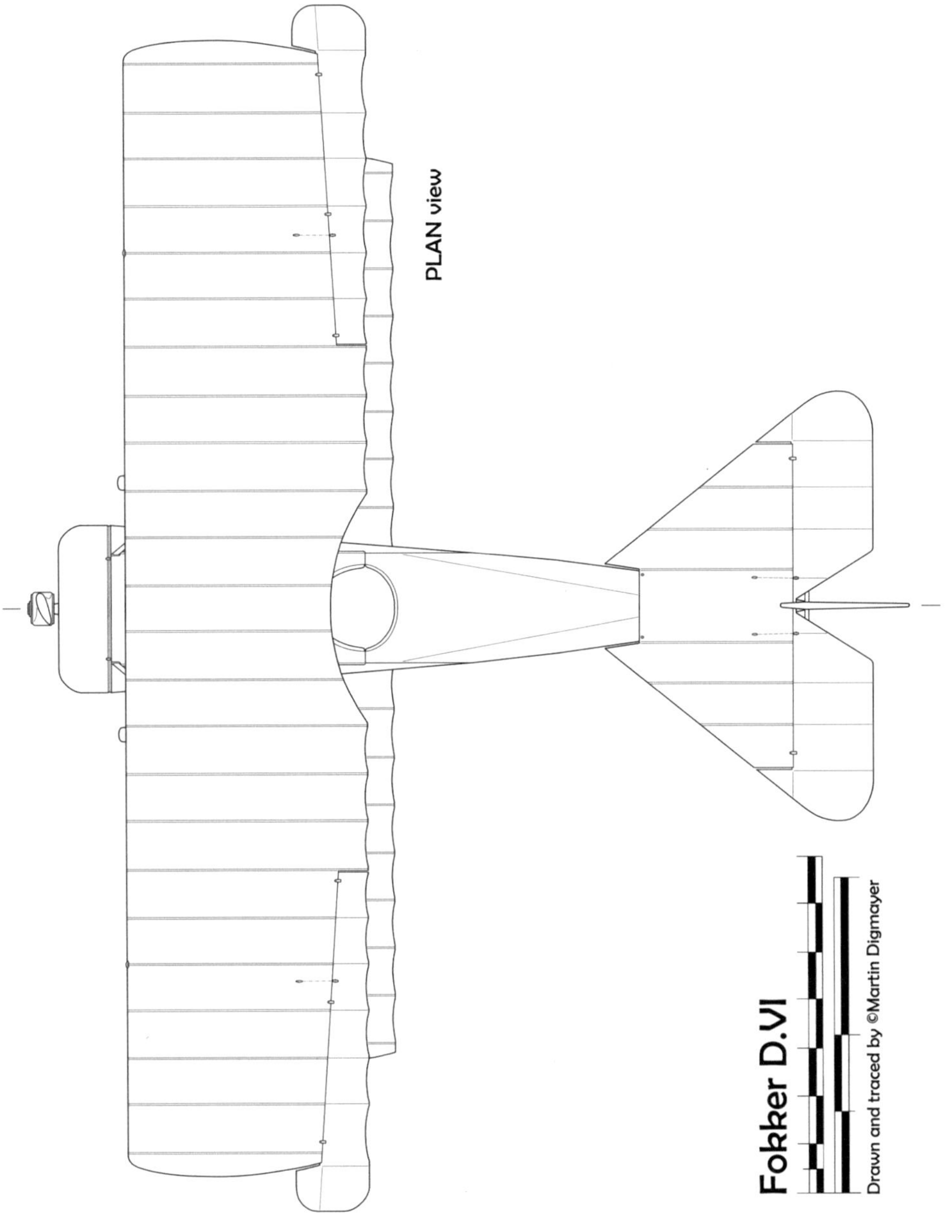
PLAN view
Fokker D.VI
Drawn and traced by ©Martin Digmayer

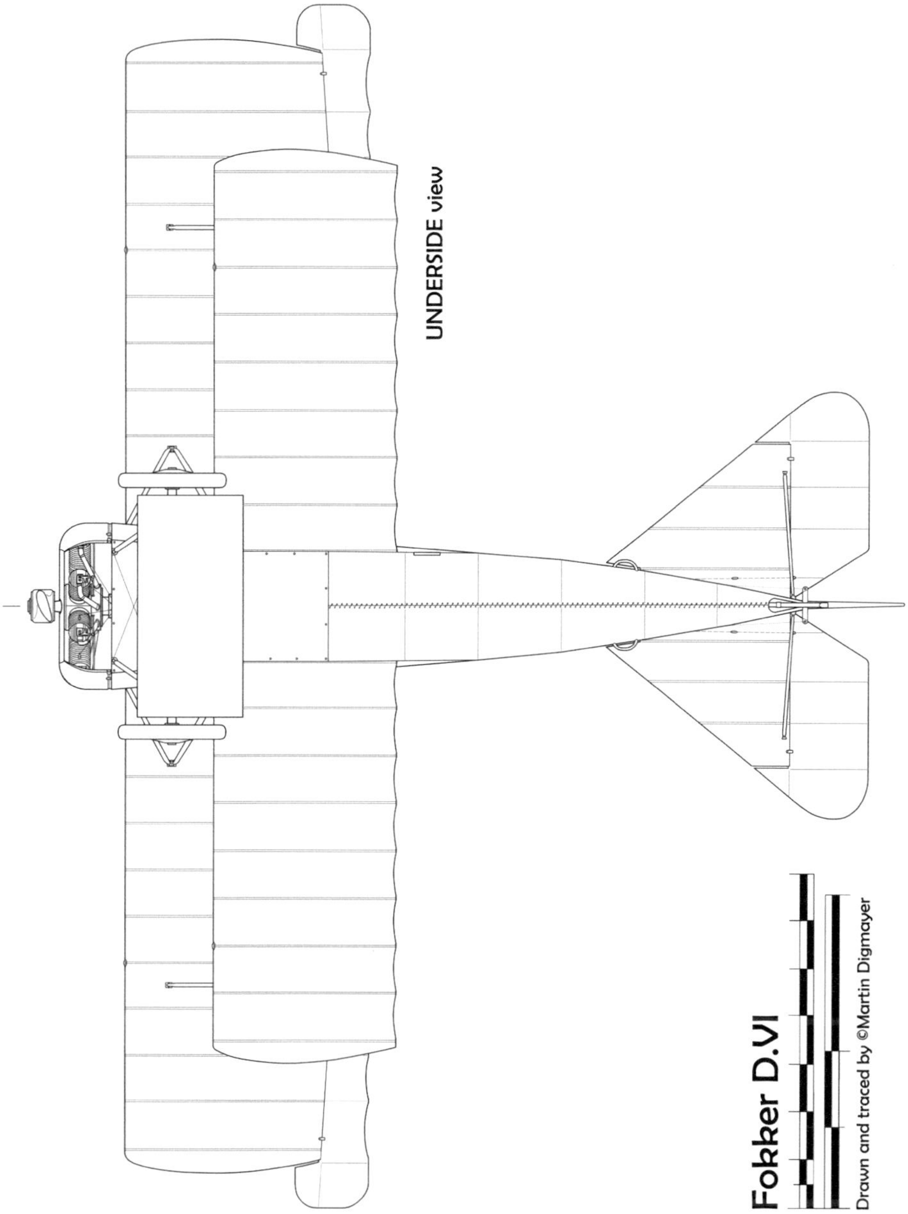
UNDERSIDE view
Fokker D.VI
Drawn and traced by ©Martin Digmayer

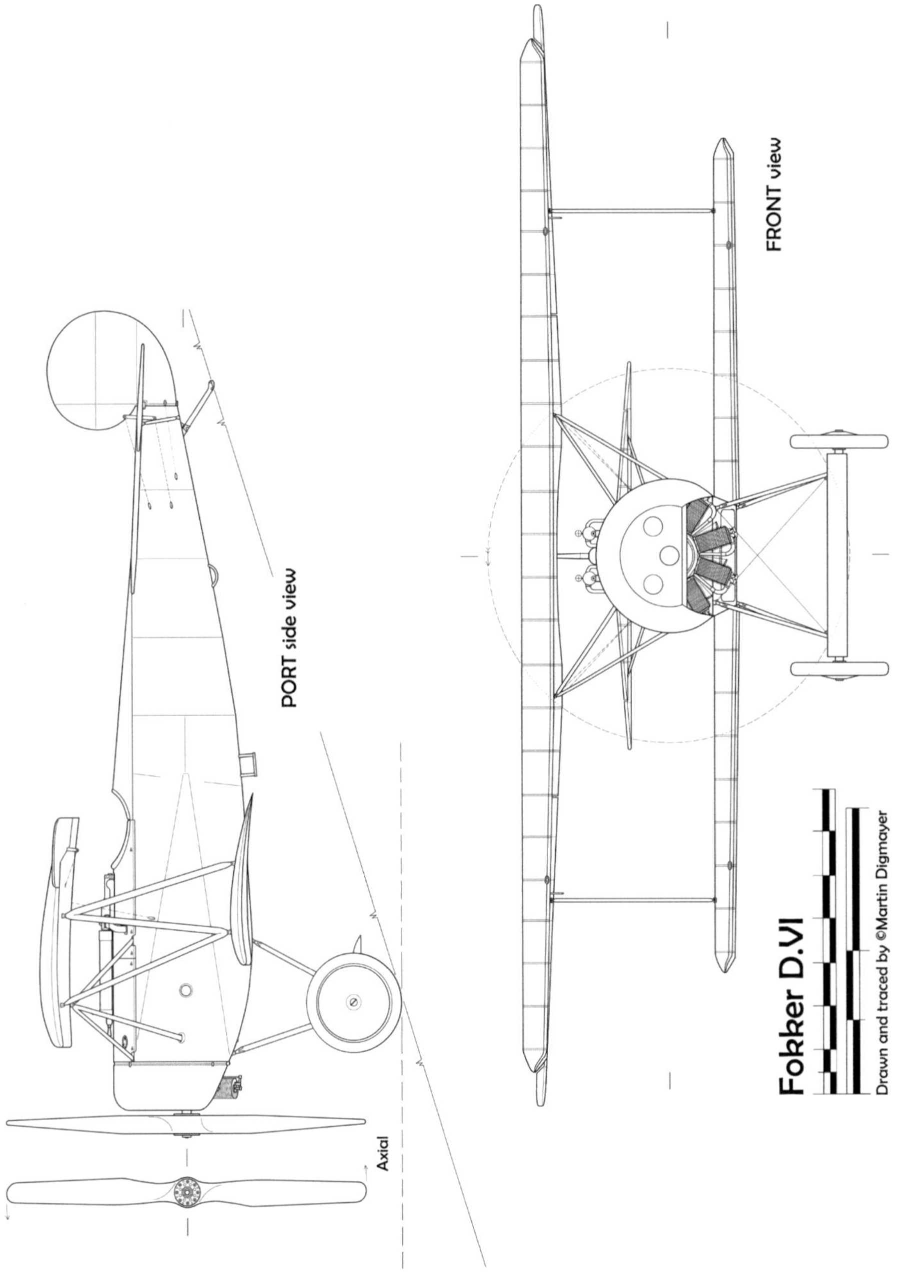
PORT side view
FRONT view
Axial
Fokker D.VI
Drawn and traced by ©Martin Digmayer

BENZ Automobile
&Flugmotoren

AXIAL
BERLIN
AXIAL-PROPELLER
sind wegen ihrer unübertreff-
lichen Leistungsfähigkeit die
erfolgreichsten und begehrtesten
Luftschrauben der Gegenwart.
Axial Propeller-Fabrik G.m.b.H.
Berlin O. 34. Warschauerstr. 58 u. 59
Telephon: Alexander 1569 u. 1588
Telegramm Adresse: Axial-Berlin

GARUDA
FELDPROPELLER
W
1914

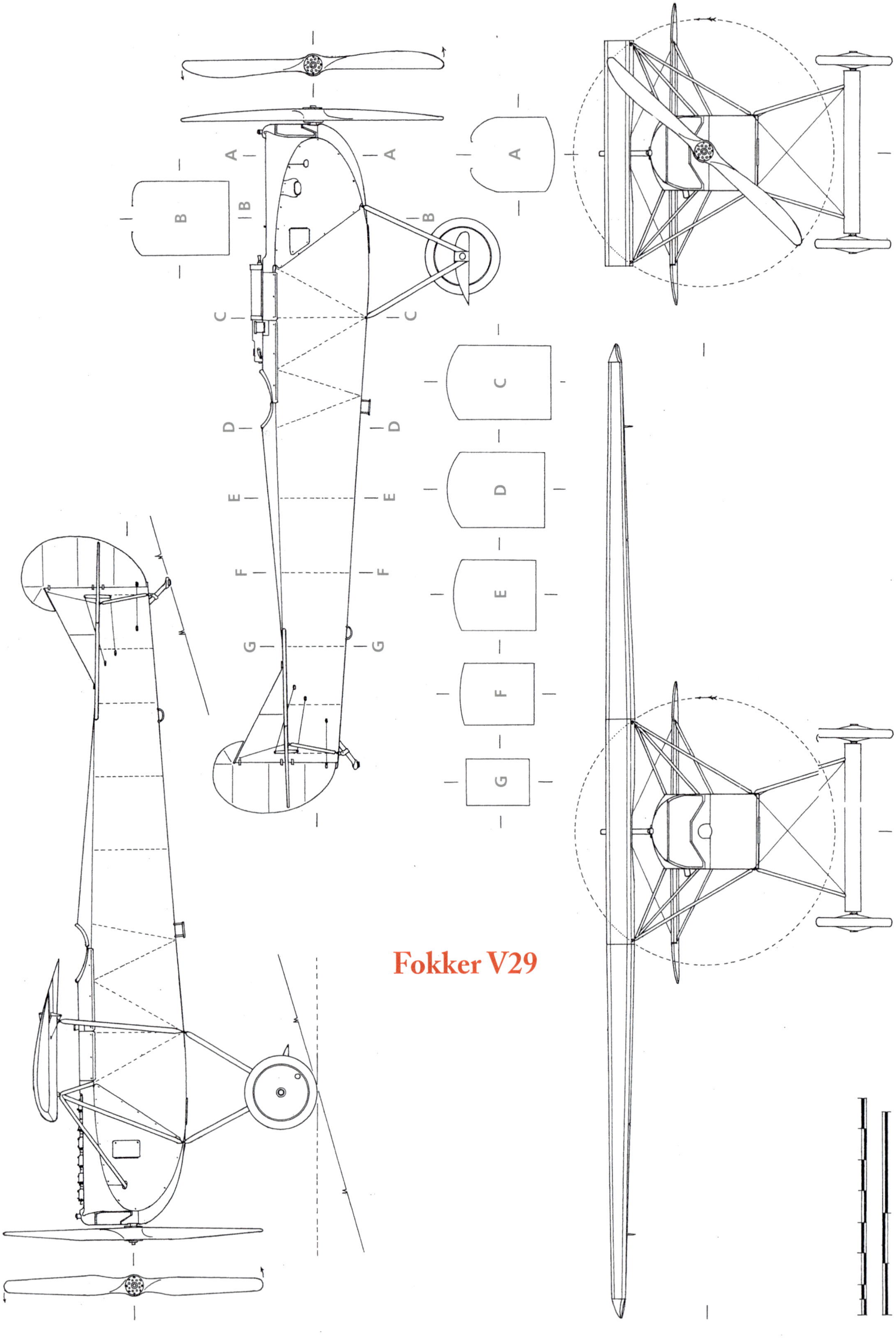

Fokker V29

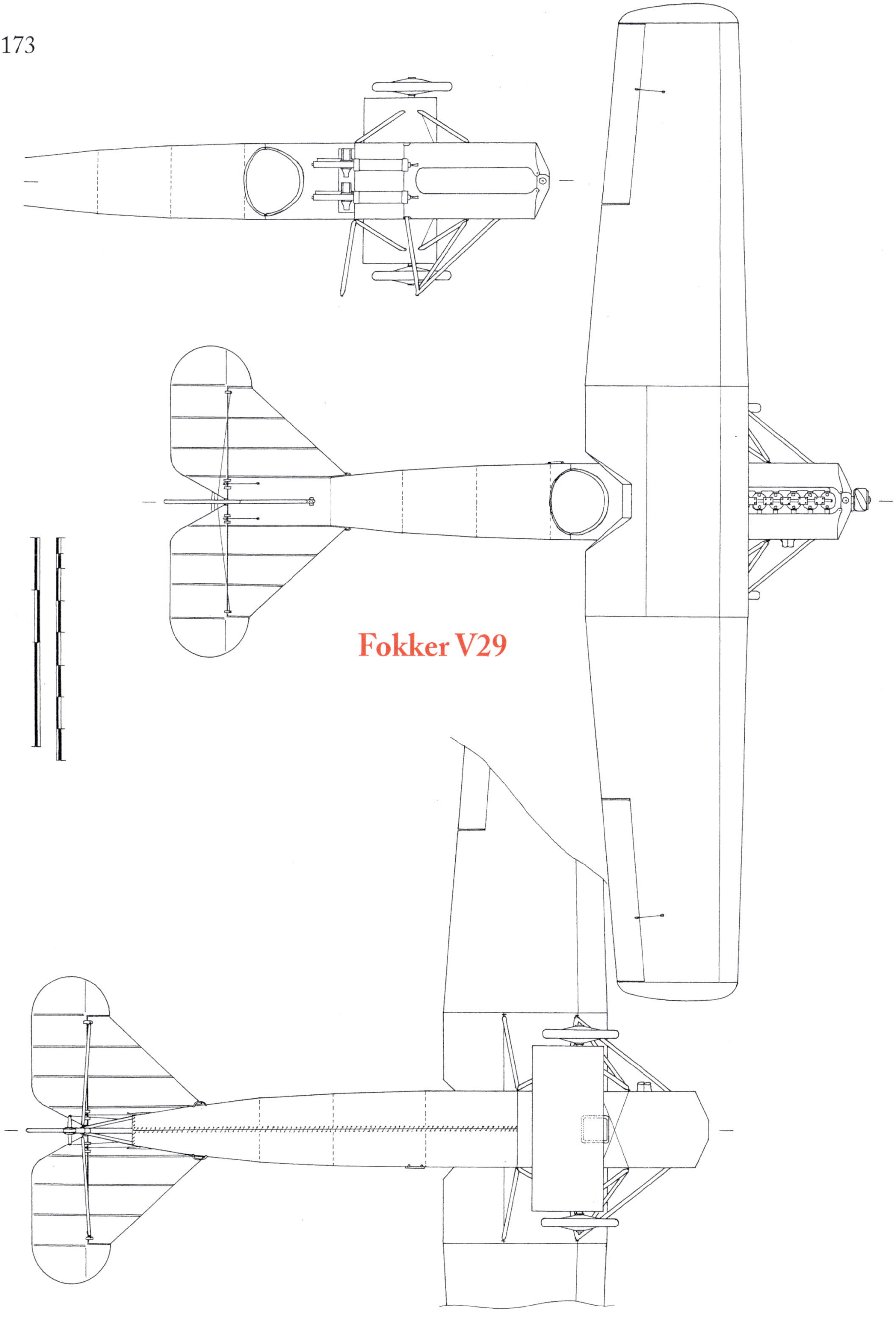
Fokker V29

Fokker Afterword

Without question Tony Fokker was the most prolific aircraft designer of WWI and one of the most successful. He had intuitive mechanical ability at a time when formal aviation engineering was in its infancy. He was at the right place at the right time, persevered greatly, and had a supportive family who helped his company survive until he was successful. He also had a nose for business and a passion for aviation. He developed the first truly successful synchronization mechanism that, when applied to his M.5K to become the world's first production fighter, transformed air combat. It also became a very profitable business for Fokker. After disappointing second-generation biplane designs, Fokker and his team developed the wooden cantilever wing with thick airfoil that made his later fighter designs so excellent. This technical breakthrough enabled Fokker to build the best all-around fighter of WWI and carried his designs to success for the next decade.

Fokker's personal qualities were more mixed. Like many successful people, he had an eccentric personality and limited interest in things other than his great enthusiasm about aviation. His cautious and conservative approach to his business finances were appropriate to his situation at the beginning, but became less appropriate as he became more successful. He had no use for formal education, another attitude that became less appropriate as aviation technology rapidly advanced and became more complex and scientific.

In business Fokker was good at finding and making deals, but paid little attention to management details. This was consistent with his general lack of personal discipline and had tragic results as Fokker aircraft, especially the Dr.I and E.V, suffered from quality control issues throughout his career. While one can argue that as owner of the company his responsibilities were high level and his employees should have done a better job on quality control, as owner of the company it ultimately was his responsibility to ensure that his production department actually *did* focus on quality control. That seems obvious, but Fokker never had the necessary personal discipline to do it.

By the early 1930s Fokker was no longer contributing to aviation technology, which, characterized by all-metal semi-monocoque structures, by then had surpassed his technical ability. Yet he is justly famous for his WWI fighters, and the Dutch company bearing his name lasted until 1996, nearly six decades after his own passing.

Below: The same aircraft was often used as a backdrop for several pilots, and this is the case here with this anonymous Fokker D.VI from *Kest* 7 in late 1918. In an almost identical photo seen earlier in this book, a *Lt.* Hoffmann poses with the fighter, while here another unit member has his picture taken with the Fokker fighter. The fuselage crosses have been modified to the thin *Balkenkreuz* typical for late-war fighters.

Made in the USA
Monee, IL
03 August 2026

58921308R00105